# PRACTICAL WELDING
## SECOND EDITION

LeRoy A. Scheck

G. C. Edmondson

McGraw-Hill

New York, New York    Columbus, Ohio    Mission Hills, California    Peoria, Illinois

# ACKNOWLEDGMENTS

The authors wish to thank:

Jacqueline Marsall and Lori Apthorp, photographers

James P. O'Neill, welding instructor, Santa Monica College

Santa Monica College, for the use of their facilities for the photographic work

Bill Vanasdale, Phoenix Welding & Supply Co.

The various manufacturers of welding equipment shown in this book.



Printed in the United States of America



Send all inquiries to:
GLENCOE/McGraw-Hill
15319 Chatsworth Street
Mission Hills, California 91345

Library of Congress Catalog Card Number: 83-81221

ISBN 0-02-829730-X

6 7 8 9 10 11 98 97 96 95 94

# CONTENTS

# PREFACE

This is a practical, hands-on text for the student seeking certification in welding. It is written in basic, easy-to-understand language and emphasizes shop experience in current welding practices rather than formal classroom theory. Exercises are designed so that the student can achieve proficiency in the basic welding methods: arc welding, MIG welding, TIG welding, oxyacetelene welding and cutting. Extensive illustrations and pictures are included to describe equipment, welding position, joint design, and the actual welding to be done in the exercises.

While the order of presentation has been carefully planned according to proven teaching methods, the book is structured so that the material may be taught as a whole or in part and in any sequence. Further, the book may be used by a student seeking certification in only one of the welding processes.

This edition has been updated to reflect advances in technique. Sections have been added to cover welding processes that were experimental or undreamed of when the book was first written. There is an entire new chapter on career opportunities in welding and prospects for the future. Additional sections at the back of the book give capsule explanations of the less common welding processes and offer information on welding inspection techniques. The glossary has been expanded and improved to cover more technical terms and the new matter included in this edition.

The text has been redesigned and reset for ease in reading and understanding. All safety notes and precautions have been highlighted.

And finally, the addition of an *Instructor's Manual* offers the instructor a useful and timesaving resource. The manual has tips on setting up and organizing work and equipment in the school welding shop, suggestions for class activities, and a complete testing program.

# WELDING

Welding, soldering, and brazing are as old as metal working: well over 4000 years. A look at objects taken from the pyramids and a walk down the Street of Smiths in any Arab city will show that methods have not changed much in 4000 years. Charcoal brazing with a mouthpowered blowpipe is still used to make the brass gadgetry that comes out of India and Arabia. Blacksmiths in the United States have not totally forgotten how to heat two pieces of iron white hot in a coal forge, dip them in powdered borax flux, and hammer them together. This is called a blacksmith or forge weld. But there are shapes that just aren't all that handy for heating and beating, so someone was always looking for a better way.

## GASES AND WELDING

In 1895 a French chemist discovered that by burning acetylene with pure oxygen instead of air the temperature of the flame was hotter than anything ever used before and, by varying the amount of oxygen, the flame could be used either to weld or cut. In less than a century more new processes have been invented for welding metals than in the preceding 4000 years.

***Acetylene.*** Acetylene is a volatile and explosive gas. It must be manufactured and stored under controlled conditions. Acetylene gas is made by dumping calcium carbide into water. The calcium carbide reacts with the water and gives off acetylene gas.

Calcium carbide looks like gray gravel, and an acetylene generator is a machine that dribbles the calcium carbide little by little into water. The generator has all sorts of safety valves and gadgetry to keep internal pressure from exceeding 15 psi (pounds per square inch), at which point the gas could explode. Acetylene is $C_2H_2$, which means that one molecule of the gas is formed by 2 atoms of carbon and 2 of hydrogen. Figure 1-1 is a simplified diagram of an acetylene generator.

As the calcium carbide reacts with water and gives off acetylene gas, a residue settles to the bottom of the generator tank. This residue is called slaked lime, a heavy white sludge which is harmless but has to be cleaned out of the bottom of the generator periodically. Sometimes farmers use it to improve their soil, if they have acid ground.

After leaving the generator, the gas is pumped into a manifold to which are attached acetylene cylinders. It takes several hours for the gas to trickle into the cylinder and be absorbed into the porous matter and acetone, which remain in even an empty cylinder unless some fool tries to blow himself up by opening a cylinder on its side or upside down. **Never open a valve on any gas cylinder unless it's upright.**

Acetylene is unstable, which is another word for explosive. In its free state it should never be allowed to reach more than 15 psi. This means that after it leaves the cylinder and is being used for welding, cutting, heating, or anything, the pressure reading on the low pressure side of the pressure regulator must never be more than 15 psi. After it is released from the cylinder, the gas becomes unstable and the molecules tend to separate from one another. As the pressure becomes higher the molecules move faster and rub against each other harder (Boyle's Law), and when the pressure reaches a certain point (which can vary according to the weather) the gas may explode violently. For a long and happy life, remember this. The only reason acetylene may be stored in the bottle at pressures higher than 15 psi is because inside the bottle it's not a compressed gas. It's an *absorbed* gas. The calcium silicate or asbestos and acetone hold the acetylene just as a bot-

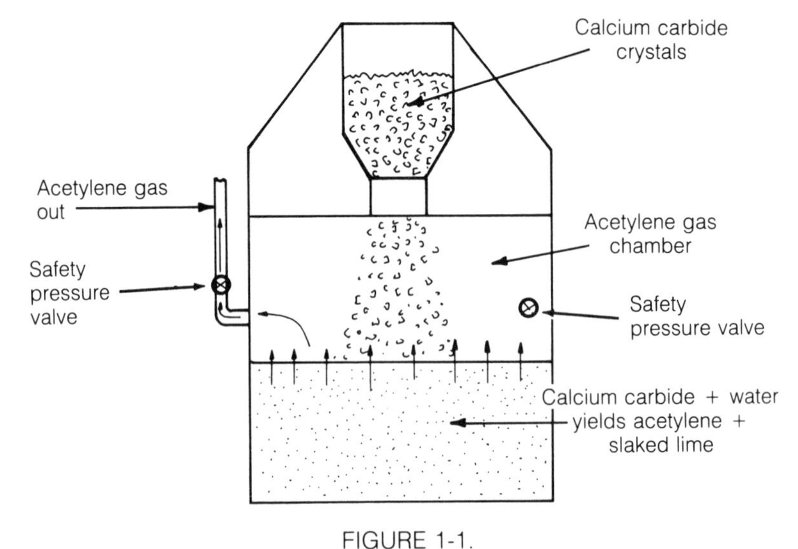


FIGURE 1-1.
Simplified diagram of an acetylene generator.

tle of root beer holds the $CO_2$ which makes all that foam when the pressure is released.

Acetylene, propane, and MAPP gases, all used in welding, are stored in liquid form. Liquid damages regulators, hoses, buildings, and people. For a long and prosperous life, never open the valve of any gas bottle lying on its side.

***Oxygen.*** About 20% of the air we breathe is oxygen, but most of it is nitrogen. Nitrogen won't burn and it cools off a flame. This is why an oxy-acetylene flame is hotter than an air-acetylene flame. An oxy-acetylene flame doesn't have 79% nitrogen flowing through it doing nothing but cooling it off and trying to put it out. This is also why, though oxygen is not flammable, it's dangerous. The burning object that smolders in air will disappear in one magnificent flash in a pure oxygen atmosphere. It was something like this that killed three astronauts.

## Liquefaction of Gases

About 1% of Earth's atmosphere is of comparatively rare gases, including helium and argon, which are used in TIG and MIG welding covered in a later section of this book. But here we're concerned with oxygen, which is produced by compression and liquefaction of air and by electrolysis of water.

A compressed and liquefied gas will boil just like any liquid. Nitrogen boils off at about 317° below zero, which makes it easy to separate from oxygen which doesn't start boiling until the liquid has warmed way up to 290° below zero. After the oxygen has been separated from nitrogen and other gases it's stored in cylinders.

## Electrolysis of Water

Another way to obtain oxygen is by electrolyzing water into hydrogen and oxygen, but it's expensive and usually done only in laboratories. Recently though, ecological buffs have despaired of ever finding a really efficient battery and have taken to designing ways to store $H_2$ and $O_2$ separated by electricity obtained from waterwheels or windmills and burning the hydrogen and oxygen again in automobiles, furnaces, or kitchen stoves: a nonpolluting fuel whose only

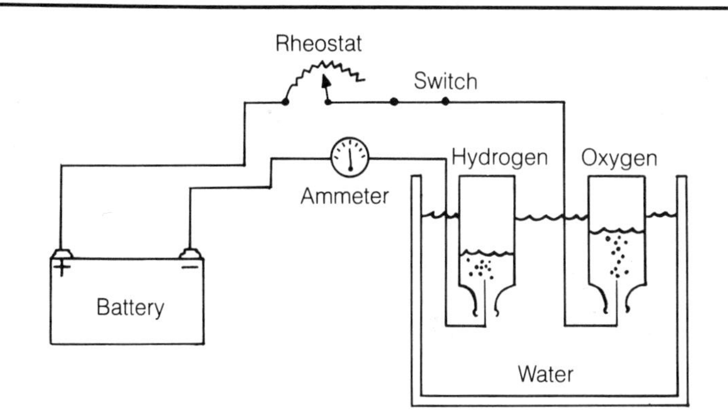


FIGURE 1-2.
Simplified diagram, showing how water is separated into oxygen and hydrogen by the use of electricity. This process is known as electrolysis.

ash is pure water. The future may see electrolysis in a big way.

H and O are the chemical symbols for hydrogen and oxygen. These elements, however, never occur in a natural state as single atoms. Hydrogen and oxygen *molecules* consist of paired *atoms*. This is why the gases are commonly written with the subscript number $_2$ ($O_2$, for example).

***Liquid Oxygen (LOX).*** Oxygen can be stored in its liquid form or in cylinders as a compressed gas. Storing oxygen in its gaseous state is a fairly simple process. The gas is pumped into cylinders, at a temperature of 70°F, to a pressure of about 2200 psi. As the temperature around a cylinder rises or falls, the pressure inside the cylinder goes up or down proportionately. (Boyle's Law again!) Thus, a bottle filled to 2200 psi at 70°, if left out in the sun or stored in a hot room, can rise to a pressure so dangerously high it can blow the cylinder's safety plug. Always store bottles in a cool place and out of the sun.

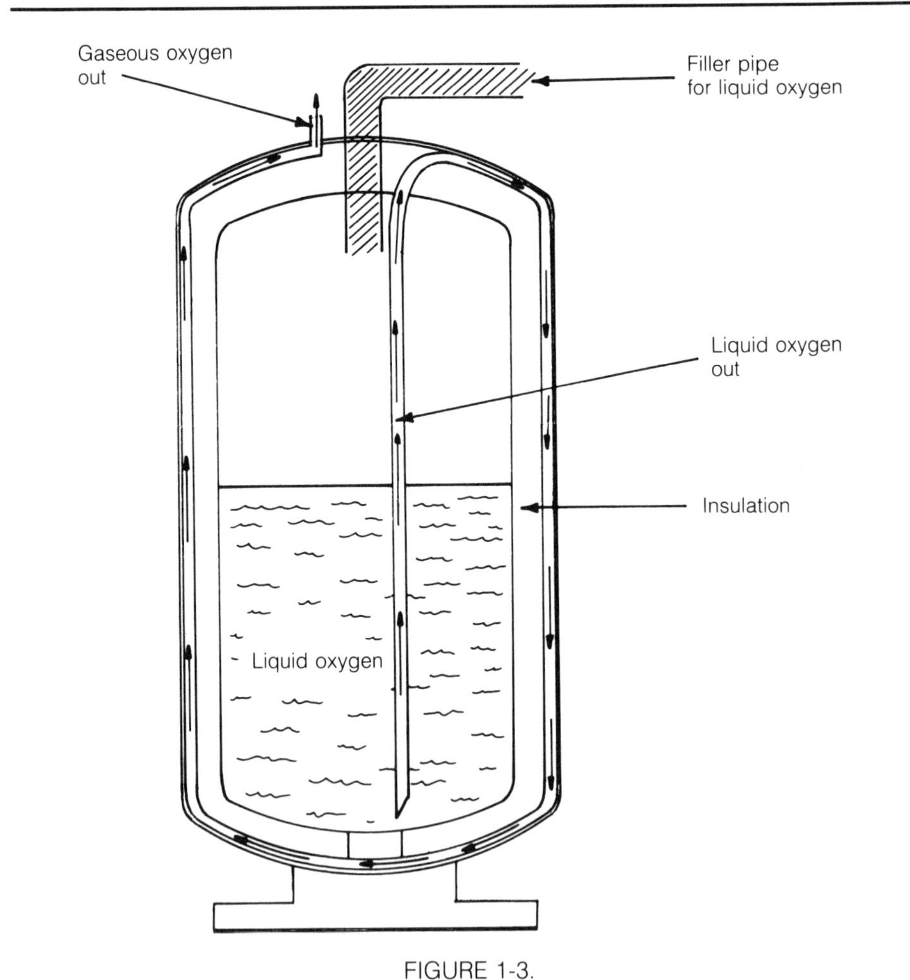


FIGURE 1-3.
Simplified drawing of a liquid oxygen container, showing how the gas is drawn off.

Liquid oxygen, called LOX by the rocket people, has nothing to do with the kind of lox eaten on a bagel. It's very stable and, since it's not stored under pressure, it does not require a thick walled container. LOX, however, is constantly warming and, as a result, changing from a liquid to a gas. It must never be kept in a container with a tight lid, or an explosion is sure to come. It's kept in a container like a huge thermos bottle, and the gas drawn off absorbs heat from the liquid. (This is the basic principle of every refrigeration system and is also another facet of our old friend, Boyle's Law.) The gas being drawn off passes through the hollow walls around the liquid and helps keep the liquid from boiling. Many large hospitals use liquid oxygen since it takes up less space and is safer than the compressed gas. A space shuttle uses more LOX in a minute than most welders will use in a lifetime.

## CONTAINERS FOR GASES

A welder must be concerned not only with the gases that unite as a powerful flame, but with the containers used for their transport and storage. These gases can be dangerous when

FIGURE 1-4.
The manifold system. Note the method of grouping the cylinders together and the safety devices to prevent tipping. All fuel gas cylinders grouped in this way must have a flash arrestor installed into the system before the gas reaches the regulators attached to the torches. A system of this type is generally used in vocational schools, or any other place where a large volume of gases may be needed. Heavy-duty, two-stage regulators are installed into the delivery lines immediately after the gas leaves the cylinders. Manifold systems are designed to store gases outside and away from the areas where they are used.

mistreated, so all containers used to store and transport them are under control of the Department of Transportation (DOT). These containers, called cylinders or bottles, are tested periodically. The date of inspection is stamped into the steel of the cylinder just below the valve assembly. They all have safety devices for the type of gas they are intended to store. The oxygen cylinder is heavier and stronger than the acetylene cylinder, and it's usually taller and thinner.

***Oxygen Cylinders.*** The pressure in a full cylinder of oxygen ($O_2$) is about 2200 pounds per square inch (psi) at 70°F. The cylinder is made of steel armor plate and must be at least 3/8 of an inch thick to comply with DOT regulations. The valve at the end of the oxygen cylinder is a little more complicated than a water faucet. It's a back seating valve, which means it's built to prevent leaks around the stem while open.

When in use the valve should be opened all the way and twisted firmly, otherwise gas can escape around the valve stem. This costs money and some unlucky day it could cost your life.

Take a look at Figure 1-6. The safety cap has a bursting disc device inside it. If the cylinder lies out in the sun or gets caught in a fire and

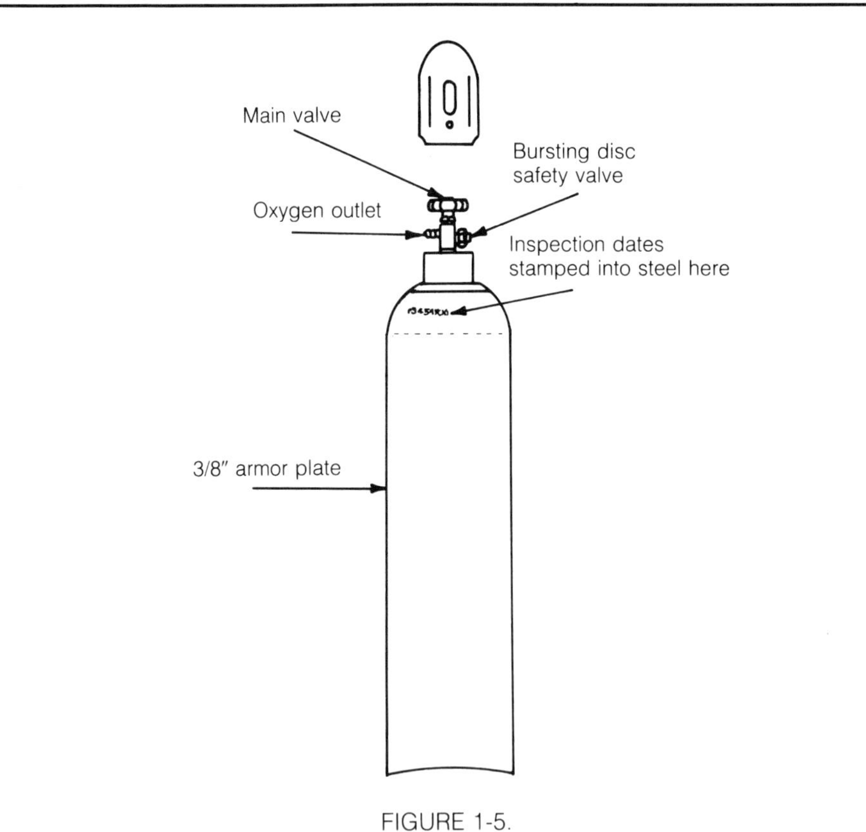


FIGURE 1-5.
An oxygen cylinder. Note the location of the safety valve (bursting disc).

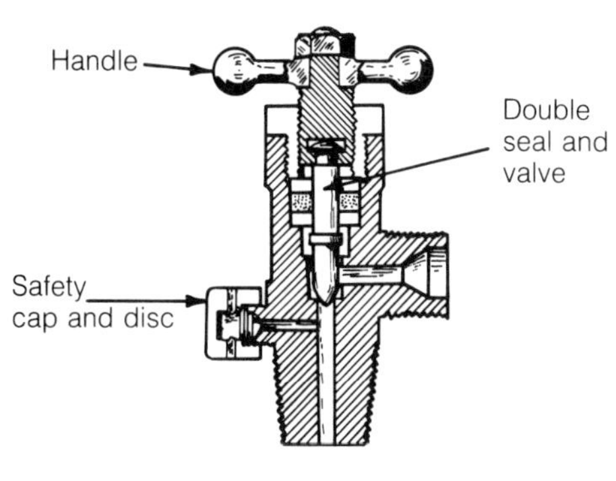


FIGURE 1-6.
The oxygen cylinder valve. These valves are also used to control other gases, such as argon, which must be stored and used at high pressures.

"Blast! I must have forgotten to chain my oxygen cylinder."
Smith Welding Equipment

the pressure rises to 3300 psi, this disc will burst and let out the gas, thus preventing a disastrous explosion. Oxygen cylinders must be handled and stored very carefully.

Cylinder pressure of 2200 pounds per square inch multiplied by the inside area of a standard oxygen bottle works out to about 5 million pounds in a single cylinder. This kind of potential energy makes dynamite look like kids' stuff.

## Safety Precautions

### WELDING GASES

1. Never remove the cylinder cap until the cylinder is ready to be used.
2. Always keep the cylinder upright whether using or storing it.
3. Fasten cylinders to a wall or on their welding cart with a chain so they can't tip over.
4. Full or empty, *never* use a cylinder as part of the ground for electric arc welding. A spark can burn a hole and the resulting explosion will make your shop look like ground zero at a nuclear explosion!
5. Don't pile anything on a cylinder or use it for a roller. It wasn't made to take pressure from the outside in.
6. When putting it into service, open the cylinder valve very slowly. Unless you wouldn't mind a faceful of broken glass don't stand in front of the gauges when opening the valve. The gas comes slamming out with a sudden jolt, but if the gauge or regulator don't blow immediately, they aren't likely to blow at all.

"I never release the adjusting screw—takes too much time. Oops!" Smith Welding Equipment

7. Keep cylinders away from excessive heat. The safety plug might keep the cylinder from exploding but several hundred cubic feet of explosive gas won't leave much of your building if there's an open flame or spark around when the plug lets go.

"When you light up, you don't miss much, do you, Jonesy?" Smith Welding Equipment

8. No matter how hard the fittings are to turn, *never* use oil or grease on a cylinder or regulator. In air, oil and grease burn. In pure oxygen they EXPLODE!

"I told you, some of the welders were using oil on their regulators." Smith Welding Equipment

***Acetylene Cylinders.*** Acetylene is stored at a much lower pressure than oxygen, so acetylene cylinders are rolled and welded, which is cheaper than the deep draw process for making oxygen bottles. The walls of an acetylene bottle are only about 3⁄16″ thick. Unlike an empty oxygen bottle, an empty acetylene cylinder is not really empty. Acetylene ($C_2H_2$) gas in its free state is extremely dangerous if stored at more than 15 psi. Fortunately, there's an easy way to get around this.

Acetylene bottles are full of something porous such as calcium silicate or asbestos. They're also about two-thirds full of acetone. This is a liquid that absorbs $C_2H_2$ just as water or soft drinks absorb $CO_2$. When you uncap a bottle of root beer, the $CO_2$ boils off and gives it a head. When you

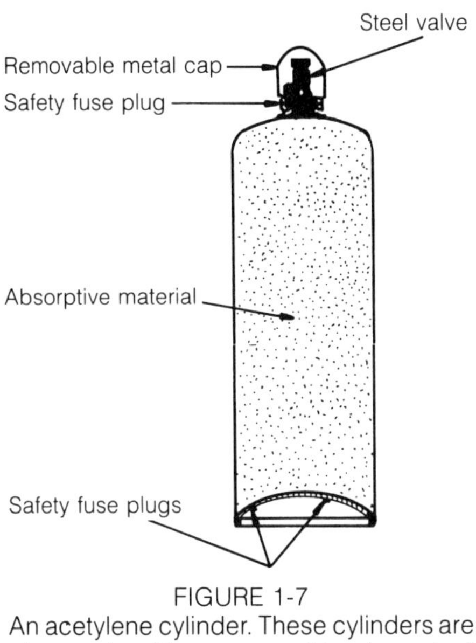

FIGURE 1-7
An acetylene cylinder. These cylinders are also made with a recessed top to protect the main cylinder valve. You will need a "T" wrench to open and close them.

open an acetylene bottle the gas boils out of the acetone in much the same way.

The safety valve in an acetylene bottle is called a fuse plug. It's an alloy that melts at 220°F, (barely higher than the temperature of boiling water) and releases the gas before it can get hot enough to explode. The welder who wants to enjoy a happy old age tries never to get a gas bottle in a place warm enough to melt these plugs, because every shop has a spark, a pilot light, or an open flame somewhere, and when this mess of gas and acetone comes foaming and roaring all over the place it's best to be in the next county.

When setting up a welding or cutting station, all bottles should be protected from excessive heat. Cutting and grinding can throw sparks much farther than one thinks. Don't let them hit your bottles.

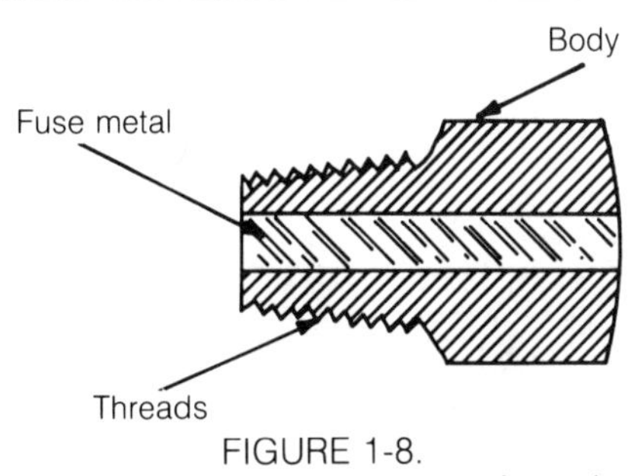

FIGURE 1-8.
The acetylene cylinder safety or fuse plug. Most cylinders will have at least three of these plugs, two at the bottom and one on the valve assembly. Some new tanks, however, do not have a fuse plug on top.

## CYLINDER REGULATORS

A regulator is the gadget between the hose and the gas bottle. It reduces the gas pressure from its high cylinder pressure down to a working pressure that will not burst the torch or blow out hoses. A regulator will usually have a couple of pressure gauges on it. The left hand gauge usually tells the pressure in the bottle, and the right hand gauge usually tells the pressure in the hose. Regulators are delicate and should not be abused. When not using the welding equipment always unscrew the "T" handle *counterclockwise* until it spins free. This relieves tension on the springs and diaphragm inside and, since it reduces hose pressure to zero, also helps prevent gas from leaking. Remember, when the valve on the cylinder is open, the regulator and gauges are under pressure. Don't bump or bang them unless you're prepared for flying shrapnel.

There are single-stage and double-stage regulators.

As the name implies, a two-stage regulator reduces the pressure in two steps. Usually the 2200 psi is brought down to about 400 psi in the first

FIGURE 1-9.
A single stage regulator.

stage, then down to the 5 to 40 psi used for welding and cutting (burning).

A two-stage regulator is nearly always used on multiple shape cutting torches where a great volume of gas is needed at very even pressure. One make is about as good as another. A two-stage regulator costs more than a single-stage, but if you're using lots of gas it'll save lots of headaches.

FIGURE 1-10.
A two-stage regulator. Notice that the left hand dial registers the high pressure of the cylinder, and the right hand dial registers the pressure of gas released to the torch from the second stage. Two-stage regulators give more accurate control of gas pressure.

## TORCHES

The mixing chamber for the gases stored in the acetylene and oxygen cylinders is the welder's torch.

There are two basic types of combination torches used for both cutting and welding: injector type, and equal pressure type. The equal pressure type is cheaper, simpler, and more widely used. Its four main parts are the body, the valves, the mixing chamber, and the tip. Usually the torch body is of brass. The valve seats are machined and will stop the flow of gas by turning them off finger tight. Pliers or a wrench will ruin the valve.

### Adjusting the Torch

The mixing chamber inside the torch body mixes the gases in proper proportions before they ignite at the tip. The tips are usually of copper alloy. They are little nozzles to direct the flame into the weld zone. Tips come in various sizes. The proper size depends on the thickness of the metal to be welded and the position in which the welding is to be done. The size of the acetylene flame does not change the temperature; only the volume of heat being produced. Only adding or reducing the amount of oxygen to the flame can raise or lower the temperature. If the work stays cold and the weld rod will not melt into the parent metal, you're probably using too small a tip. Put on a bigger one and readjust the regulators.

Popping means too large a tip, not enough pressure, or a dirty tip. Change to a smaller tip, screw the $O_2$ and $C_2H_2$ regulators clockwise and readjust the flame with the torch valves, or simply clean the tip.

One of the most annoying things for a welder is to have a torch that is continually going out of adjustment. This happens when the bottle

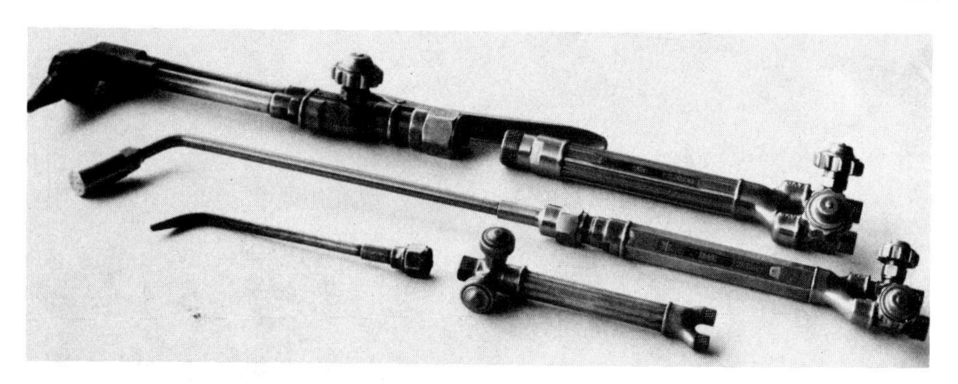

FIGURE 1-11.
The combination cutting and welding torch. The heating tip (center) may be used with these torches. A common name for the heating tip is "rosebud."

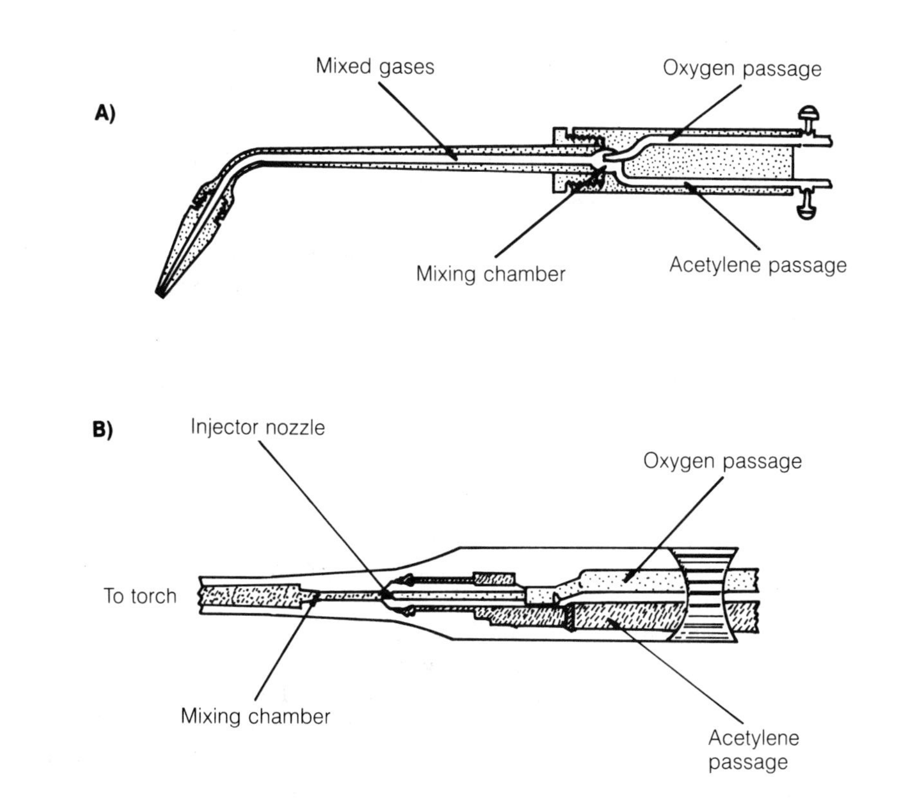


FIGURE 1-12.
(A) The equal pressure type torch. Note the way the gases are mixed prior to burning at the tip.
(B) The injector type torch. The fuel gas is drawn into the tip by a "suction" action of the passing oxygen. These torches may be used with very low acetylene pressures.

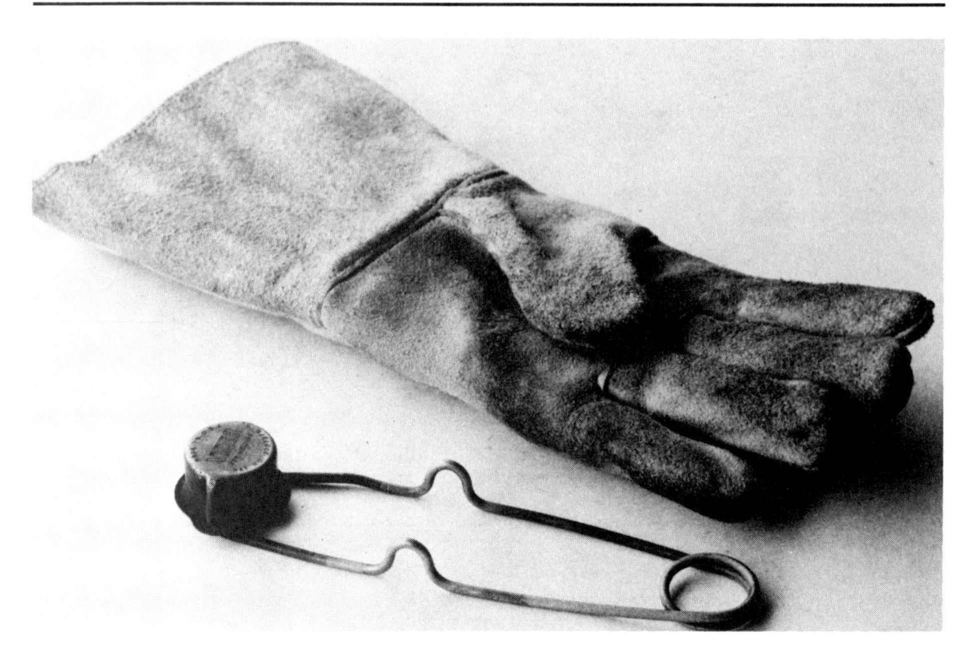

FIGURE 1-13.
Leather welding glove and a flint striker for lighting the torch.

is almost empty, or, more frequently, when the packing nuts around the torch valves get so loose the slightest touch can turn the valve. Tighten the nuts slightly with a wrench, *not pliers*. Too tight is worse than too loose, since the valve will spring back under tension of the packing each time you've gotten it just right and are letting go.

One popular make of torch has the valves out toward the end of the body next to the tip where they can be reached without turning the torch upside down. Another brand has these valves at the rear of the torch body next to the hoses. Pointing up, they rub against wrist or glove and the torch is constantly going out of adjustment. When you turn over the tip so the valves point down, they will rub on the surface below. Whatever the valves touch—the table, the work, *anything* within a yard or so—will seem to jump out and twist them just as you are reaching the ticklish part at the end of a seam.

The dangers involved in using the torch necessitate specialized equipment and procedures. These are described below.

***Gloves and Tools.*** There are other things an acetylene welder needs: A good pair of leather gloves protects the hands from intense heat and the burns that inevitably come from "popping" and splashed metal. Good leather gloves cost money and will last about one day before getting too stiff to use unless a welder gets in the habit of using pliers, vise grips, or tongs to pick up hot metal. Unless you can see water on it, *always* assume it's hot. This saves gloves, skin, and tempers. Also trips to the dispensary.

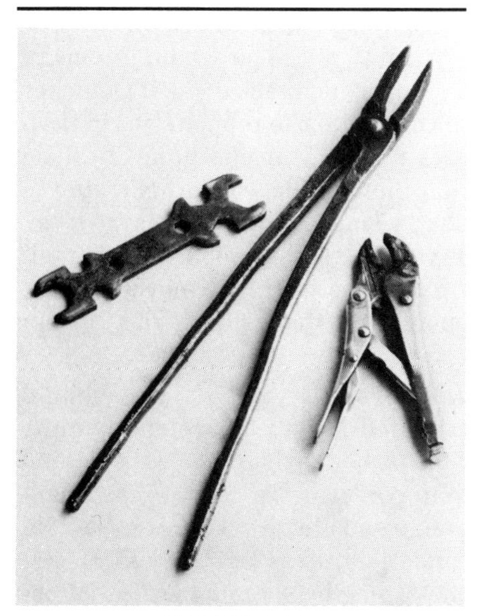

FIGURE 1-14.
Common vise grips and tongs used to handle hot metal and a torch wrench. The vise grips and tongs will give your welder's gloves a much longer life if they are used properly. Notice the different sizes on the torch wrench. These wrenches are made to fit acetylene and oxygen hose fittings as well as regulator fittings.

***Goggles.*** A good pair of welding goggles is important. They can be bought to fit over prescription glasses. Don't ever use sunglasses. Beach shades will ruin your eyes if they don't shatter and blind you first. The lenses in welding goggles are carefully compounded to protect the eyes from the brilliant flame. Use at least a Number 5 shade lens. Goggles are made with two pieces of glass in each eyepiece. The inner one is tinted and relatively expensive. Don't ever use it alone. The outer cover glass is cheap. The cover glass is the one that catches all the sparks, splashes, and spatter. Within a week or two it's pitted. Throw it away and put a new cover glass in front of your tinted lens.

This tinted lens filters out the infrared and ultraviolet rays that can cause permanent blindness. **Never weld or cut, *not even for one second,* without goggles.** It takes less than a second for the torch to pop and spray your eyes with molten steel.

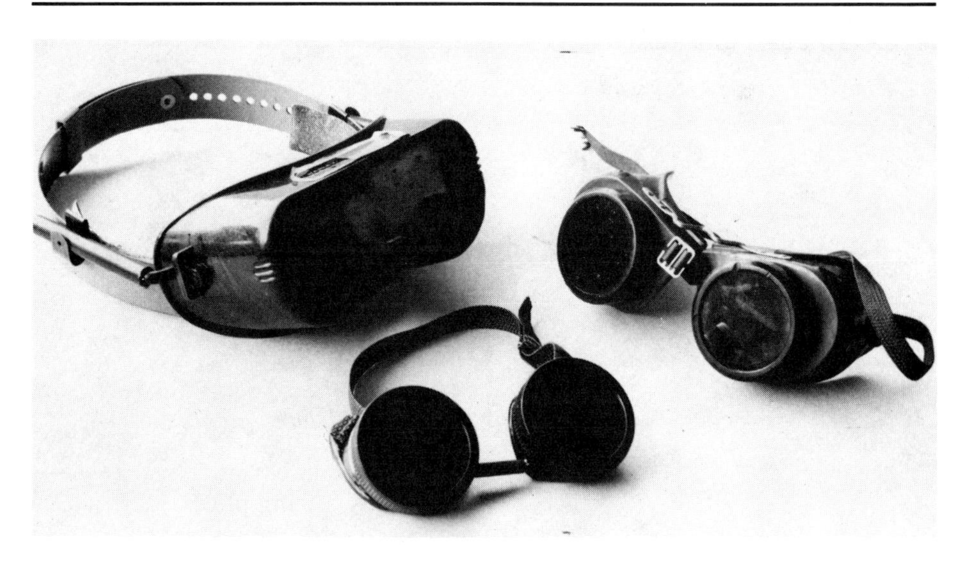

FIGURE 1-15.
Various types of welding goggles. The pair shown on the left are designed to be worn over prescription glasses.

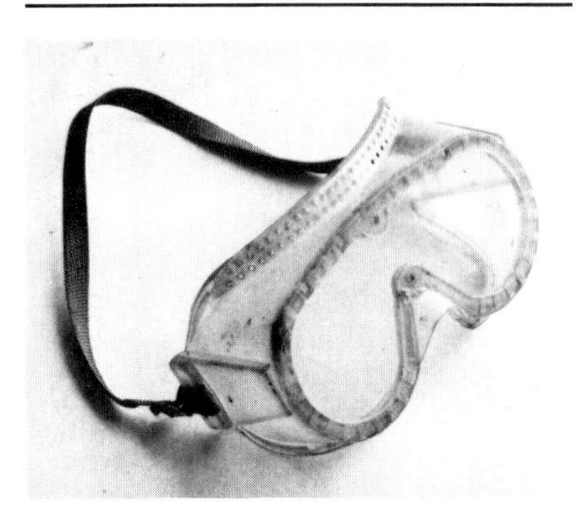

FIGURE 1-16.
Another type of goggle, used primarily for safety when grinding or drilling.

Popping is usually caused by getting the torch tip too close to the molten metal. If the torch pops while trying to light it or while adjusting the flame, you're using too large a tip or not enough pressure. Change to a smaller tip or screw the "T" handles on the regulators *clockwise* a fraction of a turn until the pressure is properly adjusted upward. Then readjust the flame with the torch valves.

Even if you're only going to make a quarter inch tack weld, it's quicker to change to the proper size tip than it is to patch up the great big hole that blew in the work from trying to burn a small flame in too large a tip. Try to do too big a job with a small tip and the work will never get hot enough for the weld to take.

***Igniters.*** Every welder should carry a flint igniter, preferably hung from a belt. Cigarette lighters and matches usually lead to bad language and blisters. There are several kinds of igniters (or strikers). Mostly they work by scraping a flint across a roughened piece of steel to strike a spark. For shops where welding is done at a fixed bench there are also gadgets with a small pilot light that burns acetylene. Usually they have a hook as part of the pilot light. When the welder hangs his torch on a hook, the gadget automatically shuts off both oxygen and acetylene.

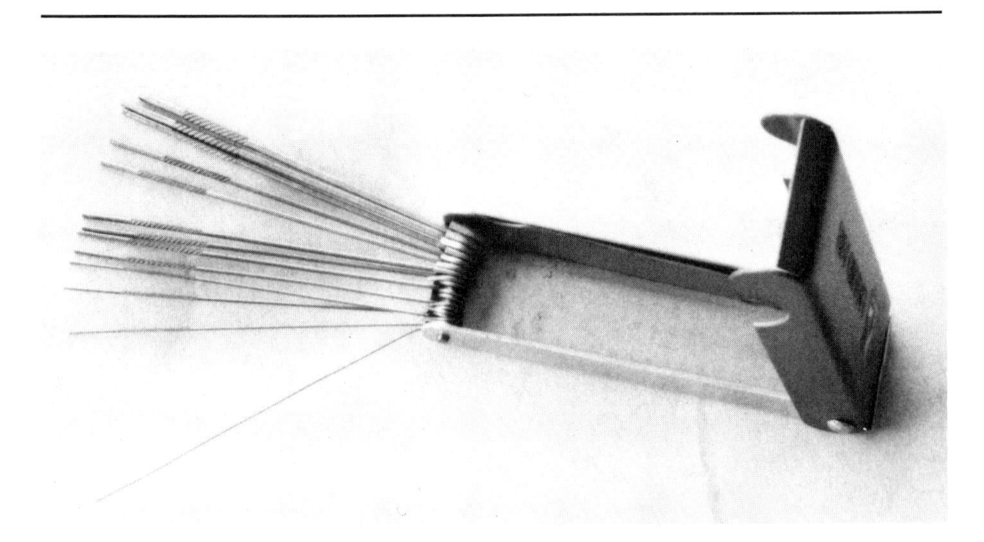

FIGURE 1-17.
A common type of tip cleaner used for both welding and cutting tips.

Unfortunately, these automatic valves usually get full of crud and fail to shut off the gas completely, leaving a torch that's either leaking or filling the shop with clouds of soot. Either way, it wastes gas. A better lighter for bench welding is the kind that shoots high voltage across an automotive spark plug when the torch is brought near.

***Tip Cleaning.*** Don't use just any old wire to clean a tip. The flame will never burn straight again. A good tip cleaner has a series of reamers, one for each size tip, to knock loose the soot and/or spatter from popped metal that accumulates inside. The tip needs cleaning whenever the flame starts hissing or getting fuzzy and crooked. Another indication is when the torch starts popping. This happens because the built-up crud inside the tip glows white hot and sets fire to the mixed gases before they can get out of the tip. It's difficult to make a good weld with a popping, dirty tip.

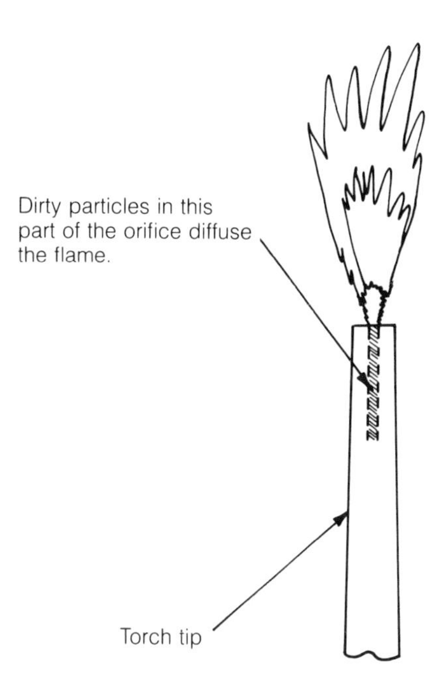


FIGURE 1-18.
How a flame may look because of a dirty tip. Notice the area inside the tip where dirt will collect.

## Safety Precautions

### OXY-ACETYLENE WELDING

Your instructor will periodically review safety precautions and give written tests. Remember, a safe worker is a good worker. Welding gases are more powerful than dynamite or TNT, and easier to set off. Important things to remember in acetylene welding are:

1. Wear goggles. The correct shade is important. Welders will use a darker glass at night than in daylight. This is because bright daylight closes down the iris. An eye in sunlight does not have to close down so far when looking at a welding flame as a wide-opened iris in the dark. A blue eyed welder should wear a darker shade than a person with brown or black eyes.
2. Be especially careful in overhead welding. Wear gloves and full leathers. Sparks *may* fly up or sideways. They *always* fall down. Several modern dances were actually invented by welders who got a big one down the neck.
3. Don't carry matches.
4. Wear leather or asbestos gloves.

5. Do not wear pants with cuffs. They catch sparks and set you afire.
6. Support all welding with fireproof materials like tile, brick, or fire brick. Make sure they're dry before using them. The slightest moisture means popping. And *never ever weld on concrete.* It won't catch fire but concrete explodes like shrapnel and even with goggles and full leathers it's like getting peppered with a shotgun. *Don't do it!*
7. Never weld or cut a closed container until it's been made safe. This is done by thoroughly ventilating the container with a continuous air flow, or by filling it with inert gas. Another way is to fill the container mostly with water, taking care that water does not contact the part actually being welded.
8. Keep grease and oil away from all cylinders containing compressed gases. Never let grease contact regulators, hoses, or torches. The results can make napalm look tame.
9. Never use welding equipment without first checking it for leaks. One safe way to find a leak is with soapy water.
10. Open oxygen and other high pressure cylinders slowly.
11. Protect cylinders from excessive heat.
12. Never stand in front of regulators when opening a cylinder unless you really enjoy an eyeful of broken glass.
13. Call all gases by their proper names.
14. Make sure cylinders are always secured so they cannot fall over.
15. Store extra cylinders separately in a well ventilated area. Keep full and empty cylinders in separate locations. "I didn't know it was loaded" will not bring anyone back to life.
16. Always mark empty cylinders.
17. Tag all leaky cylinders.
18. Keep cylinder caps on at all times, whether full or empty. If you knock one over and break off the valve you've got a rocket on your hands.
19. Before attaching a regulator, crack the valve open briefly to blow out crud that would otherwise clog the regulator and torch. Always point the blast away.
20. Always turn off the torch unless you're actually holding it.
21. Use a proper igniter to light the torch.
22. Use a wrench, never pliers, to fasten hoses and regulators.
23. A leaky regulator is a time bomb. Don't use it!

# 2 OXY-ACETYLENE WELDING

## SETTING UP THE OXY-ACETYLENE STATION

In setting up an oxy-acetylene welding station, the site must be free of all fire hazards such as wood, paper or any other flammable material. **Do not allow oil or grease to come in contact with compressed gases at any time.** Oil or grease burn in air. In pure oxygen they **explode!**

After selecting the cylinders to be used, fasten them in an upright position by chaining them to a wall, fastening them to a portable rig, or using some other means to prevent them from tipping. The cylinders are pretty solid but it doesn't take much to break off the valve. When this happens some lucky welder could be the first person on the moon without a space suit.

FIGURE 2-1.
A portable oxy-acetylene rig. Notice the safety chains used to hold the cylinders upright.

Before installing the regulators, crack the main cylinder valves to

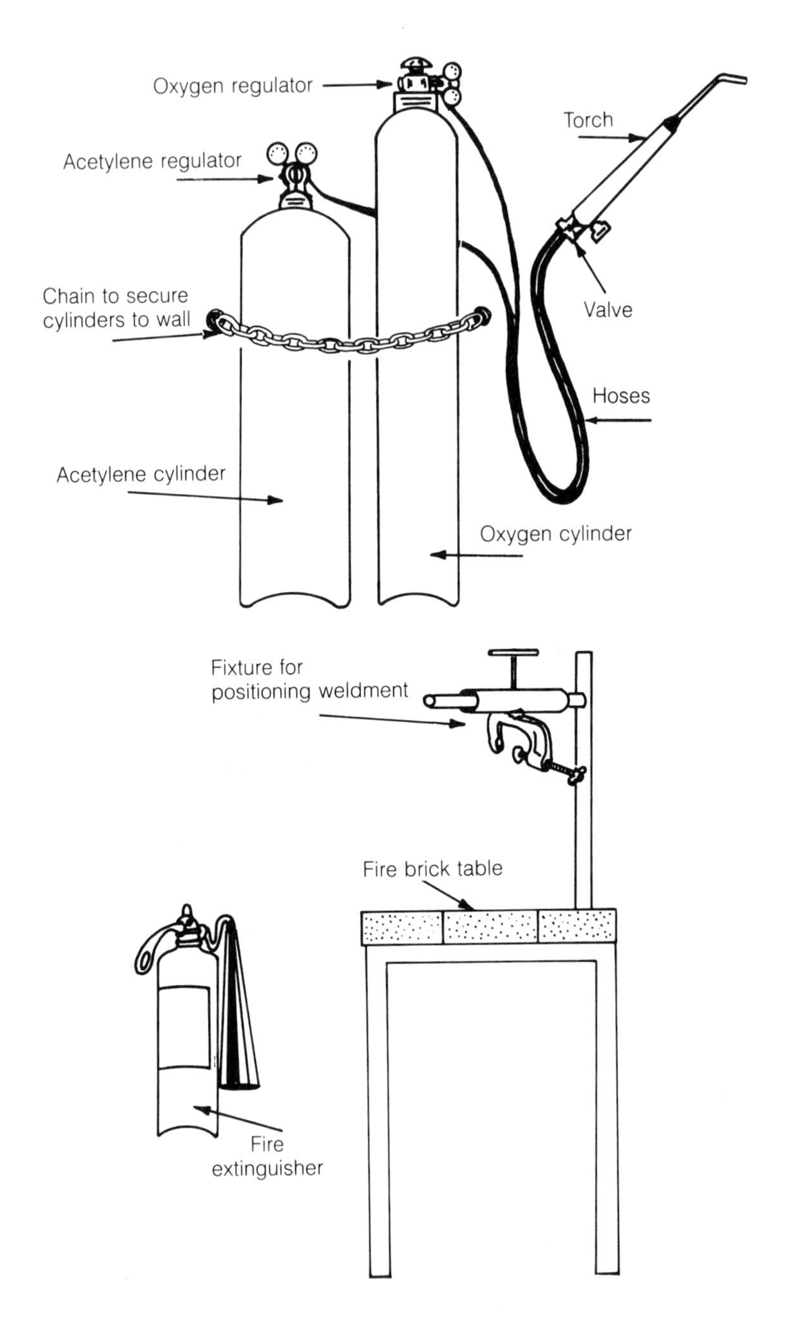


FIGURE 2-2.
Setup for an oxy-acetylene welding station.

blow out any dust or moisture. This keeps it out of the regulators and torches. Only a very short spurt of gas is necessary.

The acetylene regulator is fitted with a notched left hand thread. This is to keep some fool from putting it on the oxygen cylinder, which has right hand threads. **Never attempt to install regulators on cylinders not designed or intended for them.**

After regulators are hand screwed onto the proper cylinders, a torch wrench should be used to tighten them securely. Don't use too big a wrench or try to play Superman. The fittings are brass, which is softer than it looks. Tightening too much stretches the threads and, instead of stopping leaks it makes them worse. The next step is to connect the hoses to the regulators and tighten them snugly. Usually, the separate acetylene and oxygen hoses are joined together in a single piece. This makes for fewer snags and tangles. Attach the torch to the opposite end of the hoses and the station is ready for use. Welding hose is usually red and green or red and black. The red hose is always attached to the acetylene. Use either hose, but if you don't use the red one for acetylene the next welder may wonder what kind of an idiot hooked everything up backwards.

Check all fittings for leaks with a soap and water solution. **Never use matches or any fluid other than soap and water to test for leaks in an oxy-acetylene welding station.** Be careful when installing regulators and hoses. Damaged threads can cause a leak.

To do a good job of welding, the gas

GAS PRESSURES IN P.S.I.

| Plate Thicknesses | | Injector-Type Torch | | Equal-Pressure Torch | |
|---|---|---|---|---|---|
| GAUGE | INCHES | ACETYLENE | OXYGEN | ACETYLENE | OXYGEN |
| 32 | .010 | 5 | 5-7 | 1 | 1 |
| 28 | .016 | 5 | 7-8 | 1 | 1 |
| 26 | .019 | 5 | 7-10 | 1 | 1 |
| 22 | 1/32 | 5 | 7-18 | 2 | 2 |
| 16 | 1/16 | 5 | 8-20 | 3 | 3 |
| 13 | 3/32 | 5 | 15-20 | 4 | 4 |
| 11 | 1/8 | 5 | 12-24 | 4 | 4 |
| 8 | 3/16 | 5 | 16-25 | 5 | 5 |
| | 1/4 | 5 | 20-29 | 6 | 6 |
| | 3/8 | 5 | 24-33 | 7 | 7 |
| | 1/2 | 5 | 29-34 | 8 | 8 |
| | 5/8 | 5 | 30-40 | 9 | 9 |
| | 3/4 | 5 | 30-40 | 10 | 10 |
| | 1 | 5 | 30-42 | 12 | 12 |

FIGURE 2-3.
Chart showing suggested pressures for welding with both the equal pressure torch and the injector type.

pressure must be correct for the tip size being used on the torch body, and for the thickness of the metal being welded or brazed. The position of the metal being welded also influences the size of the tip.

## LIGHTING THE TORCH

1. Once the proper tip for the job has been selected and screwed *hand tight* onto the torch body, open the main valve on the acetylene cylinder ¼ to ½ turn. **For safety, leave the Tee wrench on the cylinder at all times while cylinder is in use. You might want to turn it off in a hurry!**
2. Standing away from in front of the gauges, open the oxygen cylinder valve *slowly*. Once it's cracked (you'll hear it and feel the jolt), open it as far as possible. This is a back-seating valve and it must be open all the way to keep gas from leaking around the valve stem.
3. Open the acetylene valve on the torch body slightly, then turn the acetylene regulator adjusting screw inward, clockwise, watching the low pressure gauge (usually on the right hand side). When it reaches the desired pressure **(never over 15 PSI!)** stop turning. Now shut off torch valve.
4. Open the oxygen torch valve and adjust the oxygen regulator to approximately double the pressure of the acetylene regulator. **Always close one torch valve before trying to set the regulator for the other!** If you don't, pressure from one leaks back up the other hose and fouls up the adjustment.
5. Now you're ready to light the torch. Until you get the hang of it, just open the acetylene valve a little way and light it. After you've been welding a while, you'll get the feel of how much to open the oxygen valve, too. A little bit of oxygen when lighting the torch stops all that smoke and those great gobs of floating soot that always land on somebody's sandwich or on the engineer's master drawing. Anyhow, use the striker, and point the torch away from yourself, or anyone, or anything else liable to burn. **That flame can shoot a yard or more!**
6. Unless you've done it a few hundred times, the torch is going to go out with a pop like a rifle shot. What you want to try for is

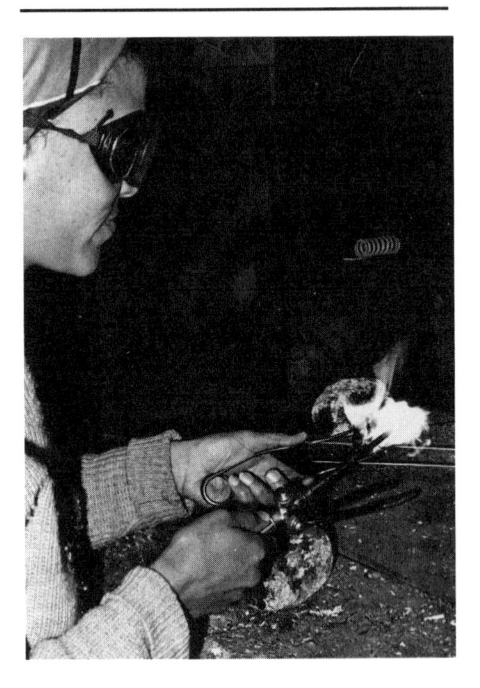

FIGURE 2-4.
Lighting the acetylene torch. Notice the direction in which the torch is directed.

a neutral flame. See the drawing (Figure 2-5) for a better idea of what this is. Keep adjusting acetylene and oxygen valves, and popping away until you get the hang of it.

## SELECTING THE CORRECT FLAME

A neutral flame has just enough oxygen to burn the acetylene. The temperature of the inner cone of a neutral flame can reach 5900°F but average temperature is about 5600°F. An oxidizing flame is hotter (up to 6300°). No matter how good a welder you are, the excess $O_2$ in this flame will burn the metal instead of welding it.

A carburizing (reducing) flame is colder than a neutral or an oxidizing flame, but the greater problem with a reducing flame is that it doesn't have enough oxygen to burn all the gas. The carbon in this unburned gas dissolves into the melted iron, making it hard and brittle. One exception is cast iron, which already has excessive carbon. Burning this out with an oxidizing flame leaves the metal less brittle. It is best, though, to braze cast iron instead of trying to weld it.

In brazing and soldering, a reducing flame is desirable since the excess $C_2H_2$ draws any oxygen out of the metals and flux, making for a clean surface and solid bonding. Stainless steel is sometimes welded with a reducing flame but, althouth these welds appear smooth, they are weak and brittle from carbon inclusions. Weld stainless steel like any other steel, with the torch adjusted to a neutral flame, which does not smoke or emit toxic fumes of any kind.

## SHUTTING DOWN THE OXY-ACETYLENE STATION

It is just as important to observe the proper procedure in closing down the station as in setting it up. These steps must be followed in order:

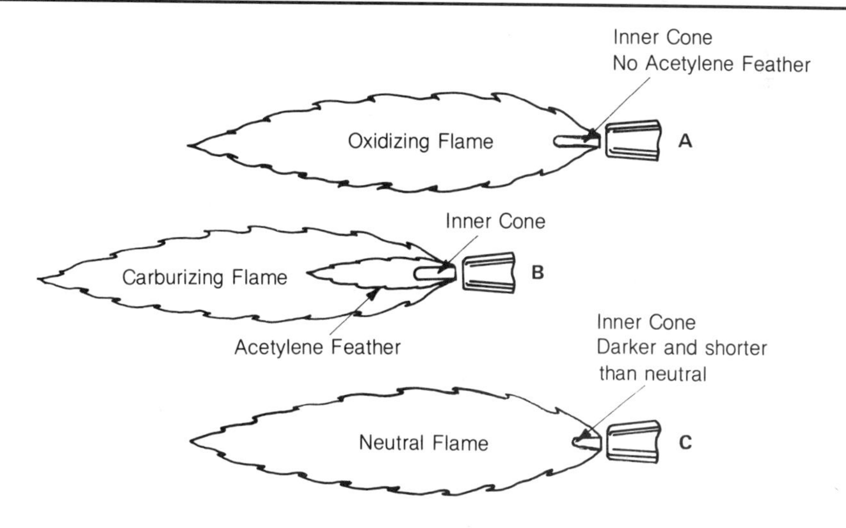


FIGURE 2-5.
Three types of acetylene flames. (A) an oxidizing flame; (B) a carburizing flame; and (C) a neutral flame.

1. When shutting down the acetylene welding station, always turn off the acetylene torch valve first. If you turn off the oxygen first the room will fill with soot and smoke and you run the risk of unkind remarks; also, a possible explosion.
2. Turn off the torch oxygen valve.
3. Turn off the acetylene cylinder valve.
4. Turn off the oxygen cylinder valve. At this point, regulators and hoses must be purged. Open both torch valves, and drain the gas from the hoses until the low pressure gauges drop to "0." Close the torch valves again, and screw the regulator adjusting screws counterclockwise until all tension is relieved from springs and diaphragms. (It is not necessary to fiddle with the regulators every time you shut down, but if the rig is to be out of use overnight or longer it helps to give the regulator springs and diaphragms a rest. They'll last longer.)

## WELDING POSITIONS AND PROCEDURES

After you've learned to light and adjust the torch you should study the four basic welding positions. Practice and learn them in the following order:

FIGURE 2-6.
Cleaning between welds. A wire brush is one of the welder's most often-used tools.

1. Flat position
2. Horizontal position
3. Vertical position
4. Overhead position.

***Flat Position.*** Now it's time to burn some metal, and the first thing to learn is called "puddling." This exercise will teach you the proper torch motions and the correct torch angles to the base metal. The direction of torch travel is opposite for right- and left-handed persons. Since most people are right-handed this book explains things their way, and a left-hander will have to reverse the direction of torch travel across the plate.

For practice, a number of mild steel plates 1/8″ thick, 3″ or 4″ wide, and 6″ long may be used. Begin on the right side and weld the short way across the plate as shown in Figure 2-9.

After the torch has been lit and adjusted to a neutral flame, hold the flame against the metal and swing the torch in short crescent motions from side to side until the metal melts. Hold the torch at about 45° to the plate and keep the axis of the torch (the flame) in line with the direction of travel. This swinging motion should make the pool of molten metal 1/4″ to 3/8″ wide. Work gradually forward across the plate, keeping the molten pool intact at all times.

If this exercise is done right there will be a small bright spot at the leading edge of the pool. This spot should follow the puddle across the plate. Keep practicing this until you get the feel of the torch and the beads are smooth and even, with penetration showing on the back side (root) of the weld. Penetration is shown by a definite burn line where the metal has tried to puddle through. This exercise is basic to acetylene welding, and will give you confidence and feel.

Plates should be welded in the same position, using welding (or filler) rod. Mild steel gas rod 1/16″ to 3/32″ can be used. When this rod is twisted into another shape it may be recognized as a coat hanger. It's easier to learn with a larger diameter rod because it won't wiggle and whip around so much while a beginner is trying to aim it into the pool of molten metal.

The motions while welding with filler rod are basically the same as those used for puddling, except that the rod is dipped into the molten pool whenever necessary to form the bead of weld across the plate. Keeping the rod tip inside the outer flame cone prevents oxidation of the hot end. Try to keep the beads as straight and even as possible. To form the bead and acquire good penetration, enough heat must be applied to the plate to melt the base metal and keep it in a mol-

FIGURE 2-7.
Proper rod and torch positions for a right-handed welder.

ten condition while the filler rod is being applied. When practicing this, keep the torch swinging from side to side and dip the rod as needed, letting the torch flame melt the filler rod enough to form the bead.

FIGURE 2-8.
Plates that have been welded by a right-handed welder.

Learners sometimes tend to let the inner cone of the welding flame dip into the molten metal. This is too close. It makes the torch pop and scatters molten metal all over the place. Keep the tip of the inner cone 1/16″ to 1/8″ from the molten pool. If you can't melt the metal with the tip held this far away from the metal you need a bigger tip. A bigger tip will also demand more gas pressure if popping is to be prevented.

After you have enough practice to make beads of uniform size and smoothness you should practice doing it backhand. This looks simple, and it is simple, once you get the hang of moving backward and doing everything in reverse. Once the backhand method has been mastered it's time to go on to other positions.

***Horizontal Position.*** In horizontal welding the face of the work is in a vertical plane and the axis of the weld is horizontal. (See Figures 2-10 and 2-11.)

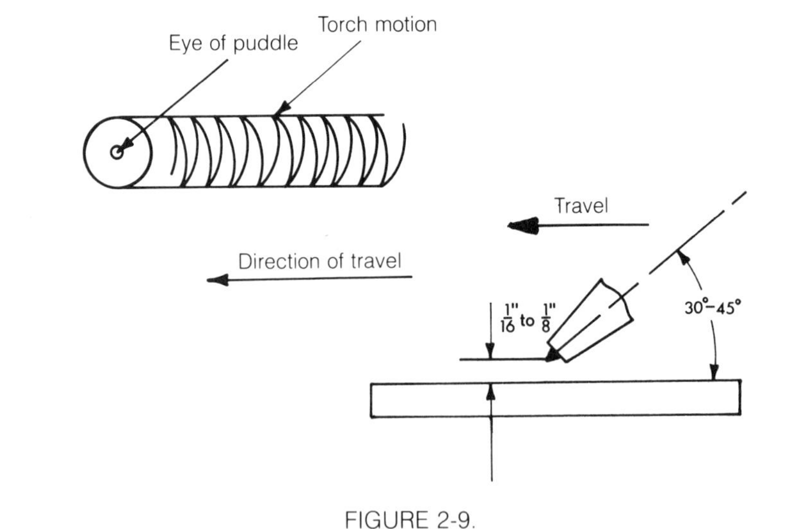


FIGURE 2-9.
Proper torch angle and motion for the puddling exercises.

This position is more difficult than flat welding because gravity is no longer cooperating to keep everything in place. The angles for both filler rod and torch are different, and these angles are critical.

The forward angle of the torch preheats the base metal ahead of the weld proper, while the upward angle of the torch uses the pressure of the gas to help push the molten metal up in place to form a uniform bead. If the torch is not held at the correct angle the metal will slop down, leaving a hollow space above the bead. This is called "undercutting," a word beginning welders will get to know quite well.

FIGURE 2-10.
Welding a plate in the horizontal position.

Figure 2-12 explains this serious "welding sin." As the weld progresses across the plate, the torch and filler rod angles need to be watched very closely to prevent undercutting. It's hard for a beginner to remember all these things at once. The most important thing to remember, however, is how to oscillate the torch. (Refer back to Figure 2-11.)

If the bottom of the bead is kept slightly ahead of the top, as shown in Figure 2-11, the bead forms a small slanted shelf on which the filler rod can be deposited, thereby making it much easier to prevent undercutting.

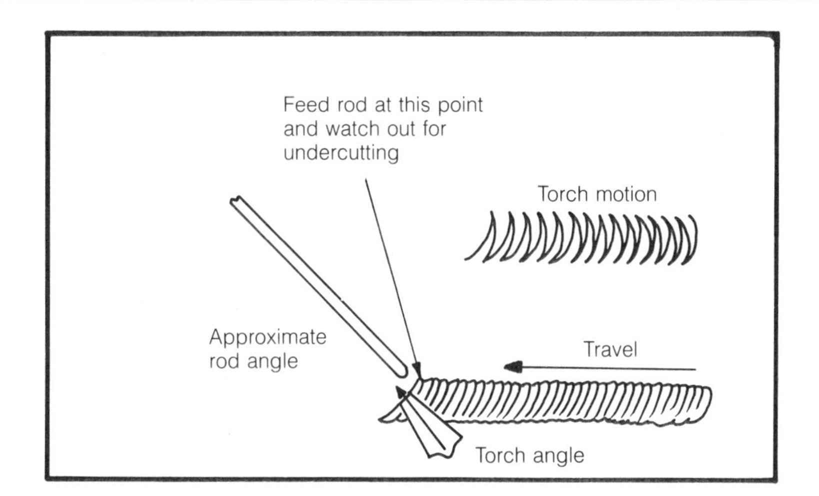


FIGURE 2-11.
Torch motion for the horizontal positions. Notice the point at which the rod is being fed into the weld.

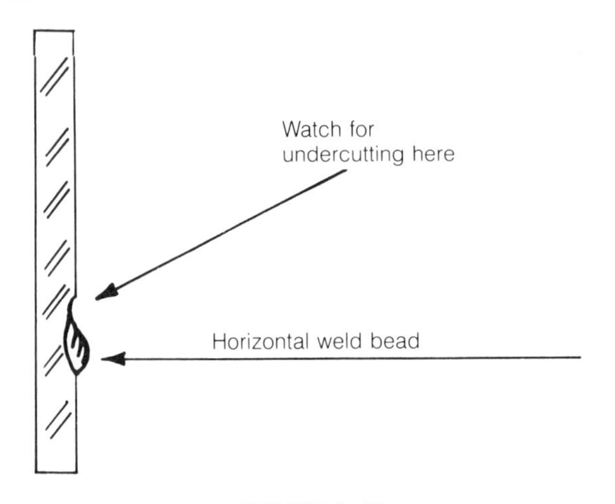


FIGURE 2-12.
The cardinal sin of welding—the undercut. The welded plate becomes thinner and causes inferior welds.

Feeding filler rod at the top of the bead and at the rearmost part of the molten pool stops undercutting, but it takes many hours of diligent practice to master this trick. It will be necessary to practice and perfect this position for most of the welding joints still to be taught. After learning horizontal welding the student may go on to the vertical position.

***Vertical Position.*** Vertical welding, as one might guess, goes up and down. Learners start vertical welds at the bottom and run up the seam. Sometimes, though, welds have to be started at the top and run down. A good welder can work either way. Place the plate to be welded as shown in Figure 2-14.

Again, this position requires a different angle for both torch and rod. The torch is held about 30° relative to the plate. This is necessary because the pressure of the burning gases helps blow and hold the molten metal in position.

Starting at the bottom, bring the base metal to a molten condition. In this position it may be better to make the bead slightly narrower than the bead you used for horizontal welding. This will help keep the weld in position. For a neat and smooth bead feed the rod sparingly. Figure 2-15 shows how to oscillate the torch.

Vertical welding with acetylene is not too difficult. A glove is clumsy and makes it harder to handle the filler rod, but it still beats burning your hand from the rising heat. After the student becomes adept at vertical welding comes the really hard position.

***Overhead Position.*** In the three previous positions there was no problem with molten metal falling on the operator. Overhead welding, however, is where the novice gets a nice hot shower.

This position is the same as flat welding, except that it's upside down, with the bead applied to the underside of the plate instead of the topside. Anything tends to fall

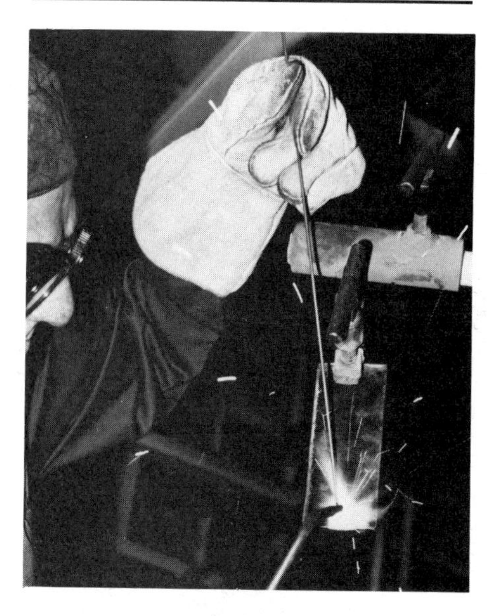

FIGURE 2-13.
Starting the vertical weld. After the plate has been brought to a molten condition the torch angle should be changed to an upward angle.

downward, and liquid metal is no exception.

Filler rod and torch angles are approximately the same as in flat or downhand welding except for being

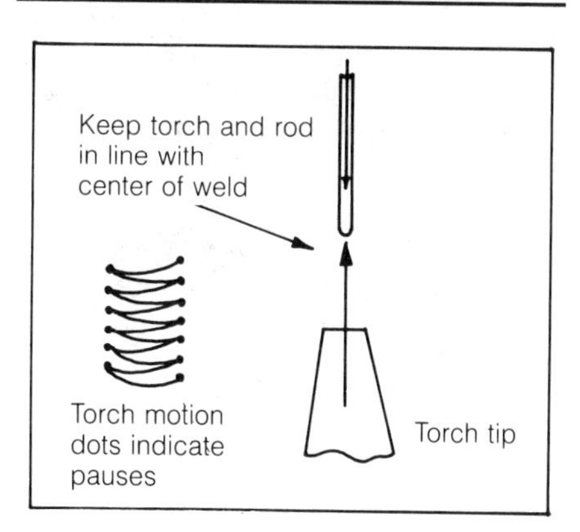


FIGURE 2-15.
Oscillation of the torch for vertical welding. The dots on the sides of the motion drawing indicate a slight pause to prevent undercutting.

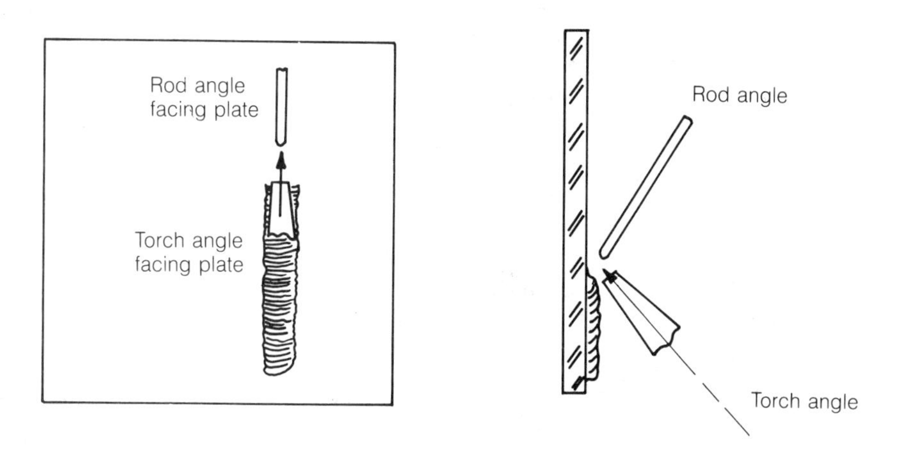

FIGURE 2-14.
Proper rod and torch angles for vertical welding.

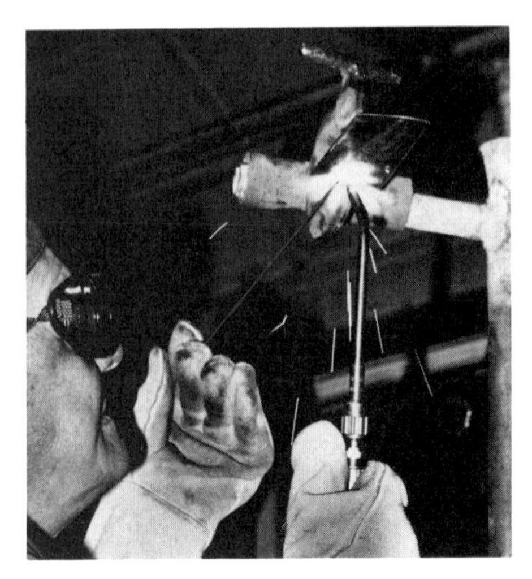

FIGURE 2-16.
Running beads overhead. Be ready to move away from possible falling metal.

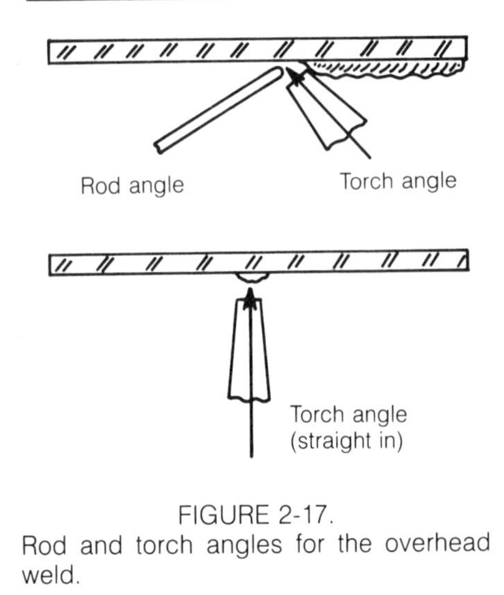


FIGURE 2-17.
Rod and torch angles for the overhead weld.

upside down. The operator should try to stand where he can escape when hot iron starts dripping. To prevent this, care must be taken not to feed too much rod. Torch angles are vital, too; the straighter the angle of the torch into the metal, the more danger there is of molten steel falling. Keep torch motion to a minimum because a large heated area falls more easily than a small one.

Until the operator gets the hang of overhead welding and can do it without burning himself, the way to keep the heated area small is to move the torch across the plate faster than usual. This is one of the positions from which you will weld for certification, and it must be learned perfectly.

## COMMON WELD JOINTS

Figure 2-18 will help you learn the proper names of the common joints. These are used to manufacture nearly everything made of metal.

***Square Butt Joint.*** The square butt joint is used to join two plates which are both lying in the same plane. This joint can be welded in any position, and is used in tanks and cylinders for storing liquids or gases.

***Tee Joints.*** The "Tee" joins one plate to another at a 90° angle. Tack it well on both sides before running a seam or shrinkage will change the two 90° angles to ones of approximately 80° and 100°.

***Lap Joints.*** The lap joint looks simple but can be difficult to weld in position, for reasons described on page 42. This joint is bulky and not so widely used as the others.

***Edge Joints.*** The edge joint is not too difficult. Sometimes it's used with thin metal to seal tank ends and can be run smoothly without rod just by melting down the parent metal. Tack weld frequently when trying to do this, because what looks like a smooth tight joint will, from heat distortion, keep opening and gaping apart just as the torch comes in contact with it.

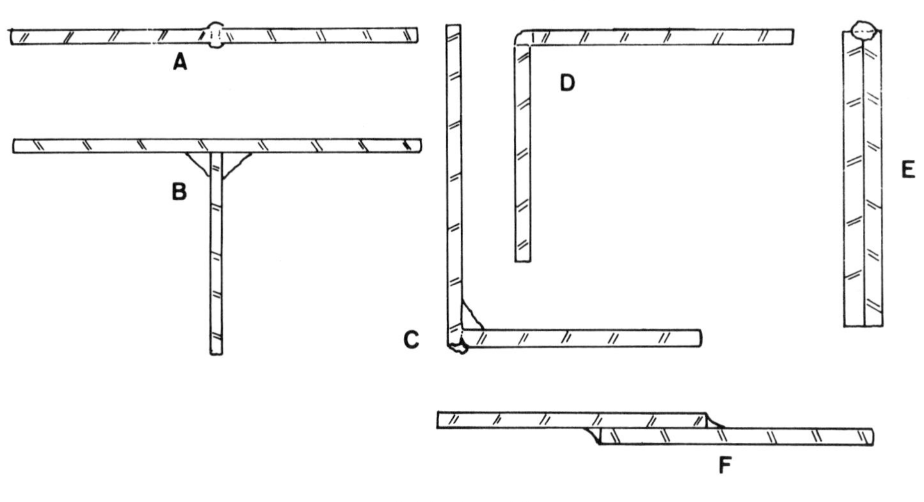


FIGURE 2-18.
Common welding joints: (A) butt joint; (B) Tee joint; (C) inside corner joint; (D) outside corner joint; (E) edge joint; (F) lap joint.

***Corner Joints.*** While corner joints are easy to weld, they are very hard to weld without warping. As the seam cools the metal shrinks and pulls inward. This warping action can be prevented by bracing, by tack welding from the opposite side, or by hammering the metal straight after it has cooled off.

## TACK WELDING

Metal always expands and distorts when heated. Tack welds are small temporary welds to hold work together and minimize distortion while the main job of welding is completed. If done properly, tack welds are small enough to blend and flow into the main weld and leave no evidence of their existence. For practicing Tee joints you should tack weld the joint at both ends, not in the middle. In time a welder learns instinctively just where to place a tack without pouring too much heat into the unbraced metal. This will usually be along some edge or corner.

In the lighter pieces that you will weld here, the tacks need not be more than ½″ in length. By keeping the tack weld small, the danger of pulling the plates to be welded out of position will be lessened. On very important work and on large weldments, tack welds must be very carefully applied so the work is not distorted in any way.

## TEE JOINTS

After the student has learned the four basic welding positions, he or she will be able to place the rod wherever it is needed. These exercises at welding beads in all positions before actually going to work are very important since they give the beginner confidence and feel. No amount of listening or reading can give anyone the latter. Feel only comes from hours of practice.

For practice welding Tee joints try ⅛″ thick mild steel. This practice gives you a chance to learn how to avoid undercutting the vertical plate and also teaches you the required torch motions to make a sound weld.

***Flat Position.*** Place the joint to be welded on the fire brick tabletop with the face of the joint toward the operator. When lighting the torch, increase the oxygen so the flame is slightly oxidized. This is necessary because the gases must exert a greater force than is normally needed in order to penetrate into the root of the weld. The exact torch angle for this weld will vary with the type of

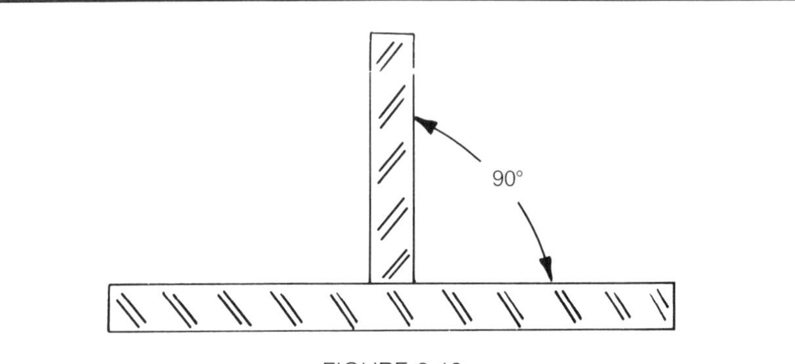


FIGURE 2-19.
How the Tee joint is formed.

FIGURE 2-20.
Welding a Tee joint in the flat position.

metal being welded, but it's up to the operator to make sure that both the vertical and the horizontal plates receive the same amount of heat. The vertical plate will undercut if you put too much heat on it or fail to keep the filler rod at the tip of the flame.

Keep the bottom of the weld slightly ahead of the top and allow the weld to penetrate the root before adding more filler metal. Once the student sees and understands that he or she can control the metal flow by changing the torch's angle and using the force of the burning gases, this weld will become easy. Be sure to weld several of these joints.

***Vertical Position.*** Welding a Tee joint in the vertical position is much easier than welding one in the flat position because there is less danger of undercutting. However, it is very important for you to watch the edges of the weld here. The dots along the edges of the seam illustrated in Figure 2-22 show where the operator should pause to make sure the weld is being properly filled.

There are no set rules to tell you what angle to direct the torch and rod into the Tee joint's weld zone, but if Figure 2-22 is used as a general guide you will soon get the hang of it.

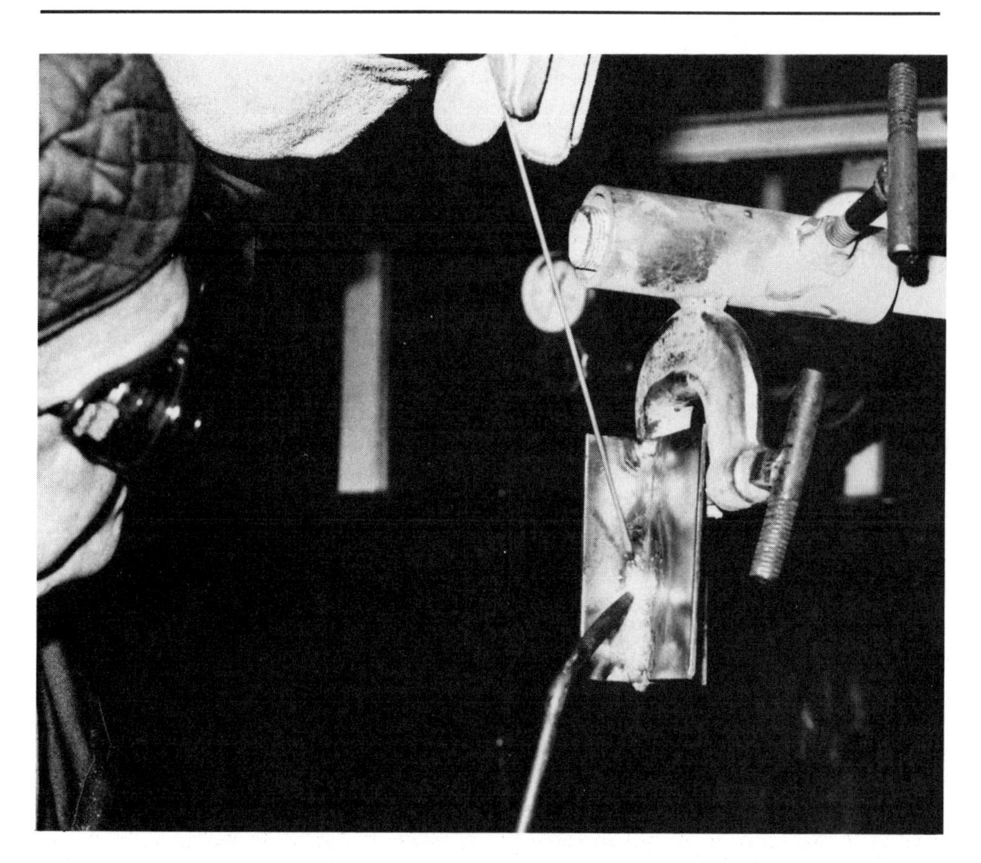

FIGURE 2-21.
Welding a Tee joint in the vertical position.

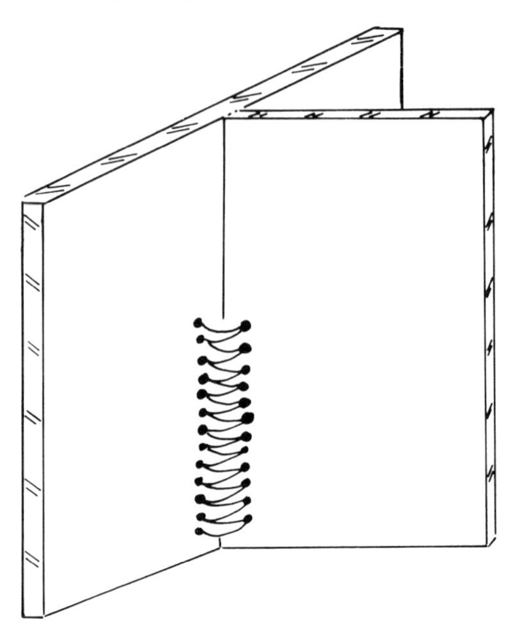

FIGURE 2-22.
Torch motions for welding the vertical Tee joint. Note that one plate is twice as big as the other. The bigger plate takes longer to heat, so the torch must spend more time on it.

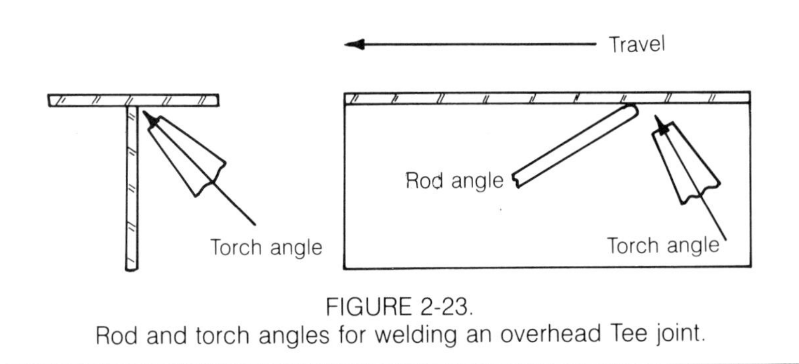


FIGURE 2-23.
Rod and torch angles for welding an overhead Tee joint.

***Overhead Position.*** Welding an overhead Tee is another matter.

The rod and torch angles must change from time to time in order to prevent undercutting. This position is not only the hardest to learn, it's also the easiest way to get burned.

Keep your eyes on the weld pool because the flow of metal depends entirely on the operator's skill to place it where and when it's needed. Make several of these joints, and experiment in the process. Change the angle of the torch more than necessary, and see what this does to penetration at the root of the weld and the appearance of the bead.

***Horizontal Position.*** After you have mastered welding a Tee joint in

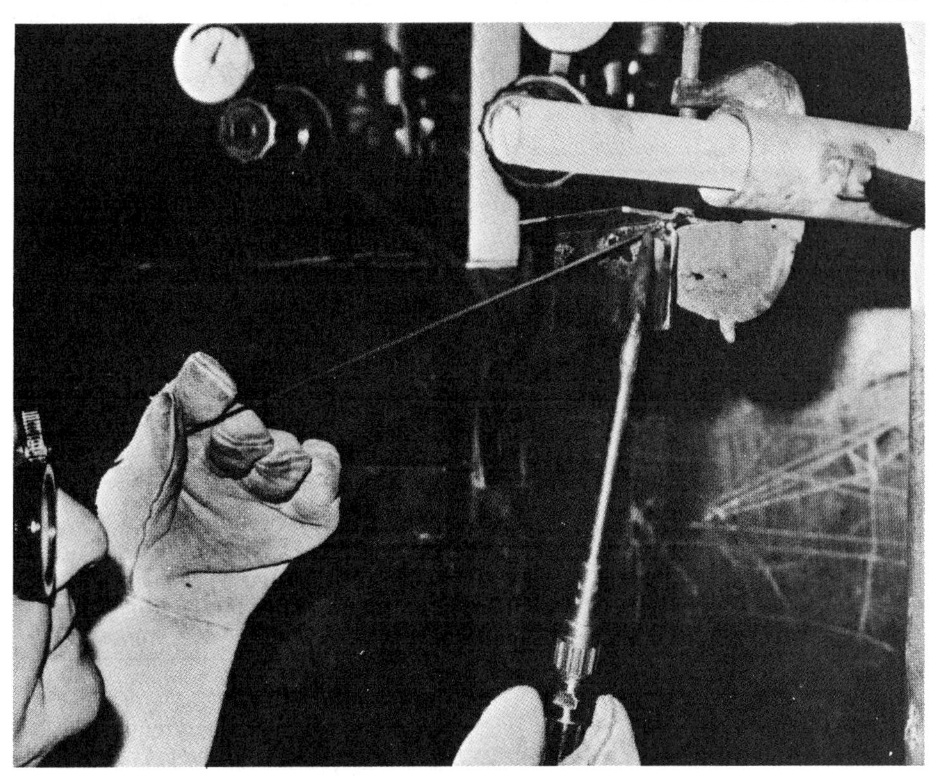

FIGURE 2-24.
Welding the overhead Tee joint. **Stay out from under!**

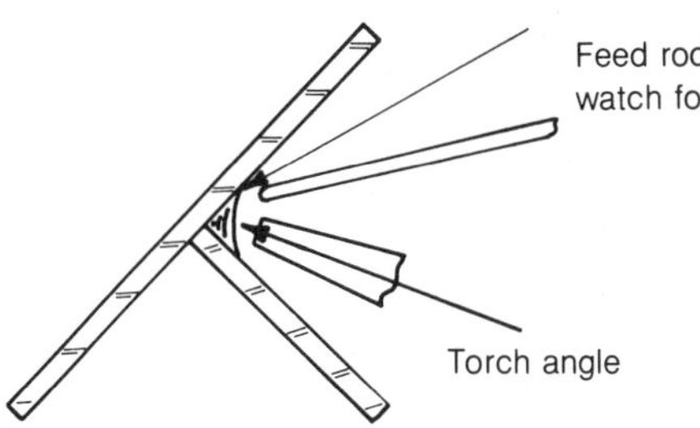


FIGURE 2-25.
Approximate rod and torch angles for the horizontal Tee joint. Watch for the cardinal sin of welding (undercutting) on the upper plate.

the overhead position to the satisfaction of your instructor, you can try welding one in the horizontal position.

Figure 2-25 shows the proper setup for welding a horizontal Tee.

In this position the welder doesn't have to worry much about falling metal, but the problem of undercutting on the upper plate is multiplied tenfold. It takes a sure hand with the torch. Again, there's no set angle for

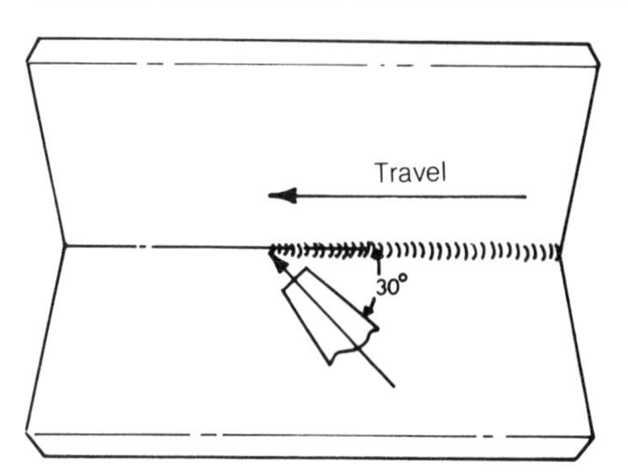


FIGURE 2-26.
Keep the torch angle in the direction of travel for the horizontal position. This angle may change from time to time to avoid undercutting.

torch or rod; however, 30° in the direction of the welding is about right. (See Figure 2-26.)

Feed the rod from the top of the pool and keep the bottom of the weld ahead of the top of the bead. Since liquids flow downward, most of the metal, aided by burning gases, will stay wherever intended. You will have more trouble learning this joint than any of the other three Tee positions. Many welders find the backhand position easier for horizontal welding. This is where you must keep a sharp eye, and correct any mistakes before they have time to grow into bad habits.

Welding is not easy, and there are things that cannot be learned from books. This one is hard. Keep trying, and don't be discouraged. An instructor can explain mistakes and try to keep you from developing bad habits, but a large part of welding is "feel." Keep practicing, and you'll learn how to do it. It is not easy to acquire certification in acetylene welding. You must spend a lot of time practicing the types and positions of welding that will be used in the test.

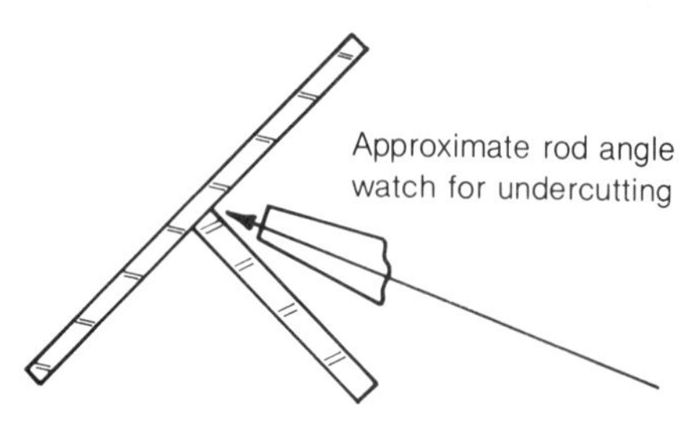


FIGURE 2-27.
The approximate torch angle for the horizontal Tee joint. This angle will probably vary from time to time and is not exact. Watch out for undercutting.

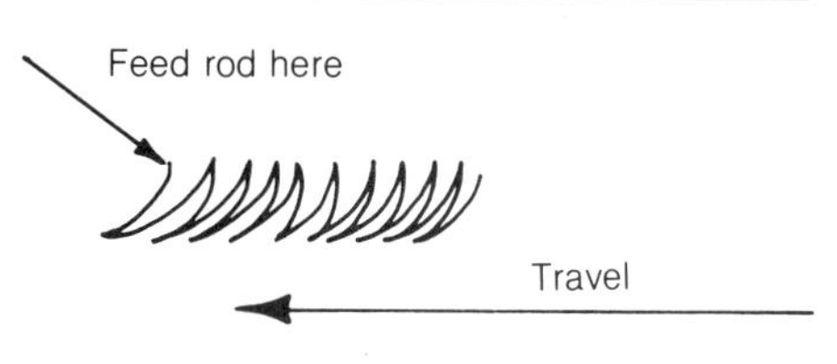


FIGURE 2-28.
Torch motion for the horizontal Tee joint. The filler rod is fed at the top of the weld to prevent undercutting of the upper plate.

## SQUARE BUTT JOINTS

After perfecting your Tee joints you should go on to the square butt joint, which must also be practiced in all positions. Figure 2-29 shows a stand for holding plates to be tack welded before running the butt seam.

***Flat Position.*** First you should practice the flat position in order to learn the action of the weld pool on the open seam of the joint. Use 1/8″ thick metal, which is heavy enough to weld without disintegrating the plate. Learning to weld thinner metals is good practice, but many times this exercise is so difficult that a beginner becomes discouraged.

When assembling the square butt joint, leave a root opening of at least

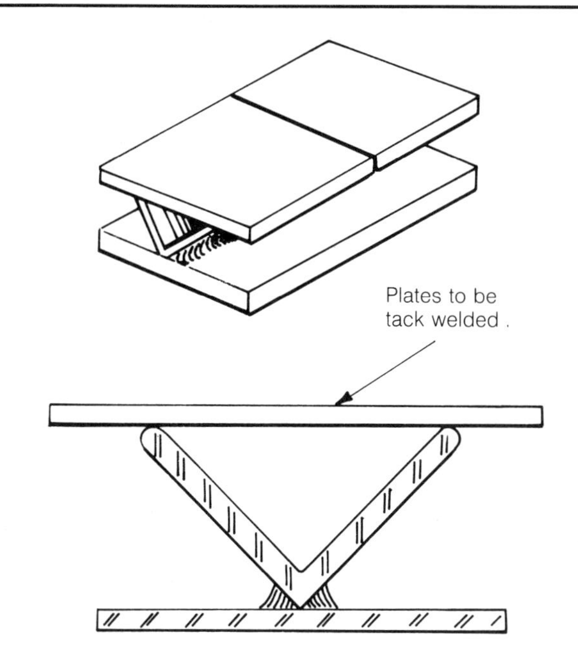


FIGURE 2-29.
How a fixture may be made to tack weld plates for butt welding.

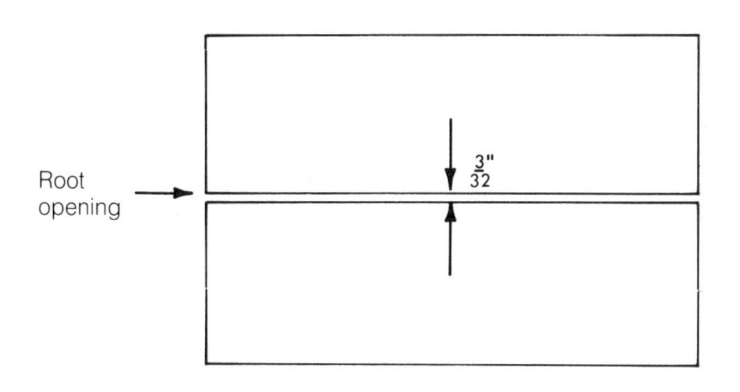


FIGURE 2-30.
Setup of plates showing the root opening for the flat position of the square butt weld.

3⁄32″. This is done by laying the plates to be welded across some surface like the face of a short piece of angle iron welded to make a jig as in Figure 2-29.

Figure 2-30 shows how to arrange the plates for practicing the square butt joint. Measure the 3⁄32″ gap by placing a piece of 3⁄32″ welding rod between them. Tack one end. Bend the plates back into shape with the proper gap again, and then tack the other end (see Figure 2-31).

After the plates are tack welded together, place them bridging across two pieces of firebrick so the back side of the weld will not be against the table top. This helps prevent penetration of the weld along the open root.

When starting this weld, be sure the weld pool is wide enough to assure complete penetration. If penetration is to be complete, great care must be taken to heat both plates equally. A small hole will appear in the open root. This is the "keyhole." This keyhole (see Figure 2-32) must stay the same size throughout the length of the weld, with the heat and amount of filler rod balanced out. In this application, "balanced out" means keeping the keyhole even; if the keyhole becomes too large, rod must be fed into the weld pool faster. This will prevent excessive penetration, because the cold rod will cool the weld pool.

After you become proficient in getting complete penetration with 1/8" metal, you may go on to a heavier plate, usually 3/16" thick. Bevel the edge to be welded on each plate 30°. Thicker metal requires a larger torch tip since thick metal soaks up more heat. (See Figure 2-33.) Beveling, or chamfering, may be done with a sander, disc grinder, or any abrasive wheel setup available.

Don't try to fill the entire groove in a single pass. It's too hard to get complete penetration. The first pass is called the root pass and must be done very carefully. After the root pass the second or cover pass is much easier.

The cover pass should fill the groove about 1/16" above the level of the base metal and should not show signs of undercutting along the edges. After the student has learned this to perfection he or she may go on to 1/4" plate; this is usually the plate thickness used for acetylene certification.

The student will soon note how much more heat is needed to weld heavier plates. An uncomfortable welder cannot do a good job, so give some thought to cooling and ventilation before starting.

In preparing 1/4" plates, use metal that is 8" long and 4" wide. Bevel the plates at the same 30° angle used for 3/16" plates. This will give you practice with plates of the same size and thickness as those used in the certification test.

Weld your first set of 1/4" plates in

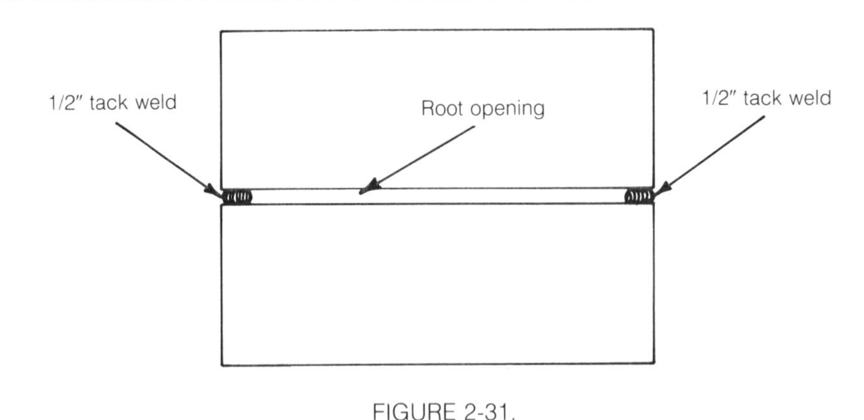


FIGURE 2-31.
How the plates are tack welded for the square weld in the flat position.

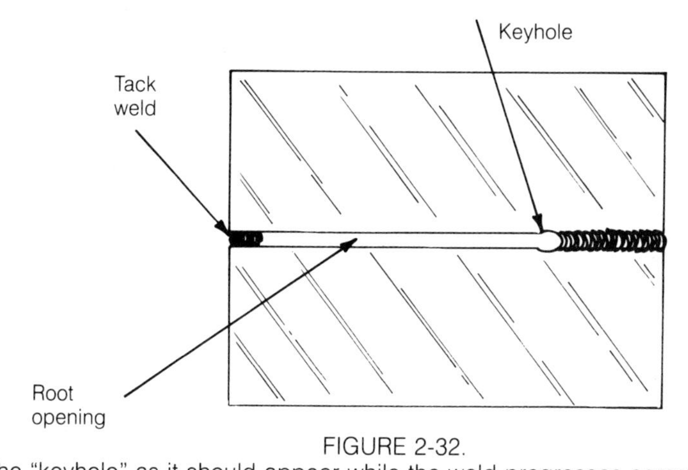


FIGURE 2-32.
The "keyhole" as it should appear while the weld progresses across the plate.

the flat position. This will give you an opportunity to determine correct tip size and proper torch motions for this gauge (thickness) of metal. Again, the keyhole will have to be established to assure perfect penetration. Torch and rod angles cannot be given because they will change in order to assure the needed penetration. **Do not try to fill the entire groove in a single pass.** Doing so will plug the weld and prevent full penetration. On the first or root pass, concern yourself with obtaining penetration, and make sure both plates are being joined into one solid piece of metal as the weld progresses. Be sure the axis of the torch tip is centered in the direct line of progression. If the torch tilts toward one plate or the other, the plate receiving the most heat will get penetration while the other will not. This is why both plates must always receive an equal amount of heat.

After welding two or three plates, test the welds for penetration and fusion of filler rod to base metal. Do this by cutting strips 1½″ wide crossways to the weld and bending them 180° across the weld. There should be no sign of separation anywhere in the weld, neither in root nor face bends. See Figures 2-35 and 2-36.

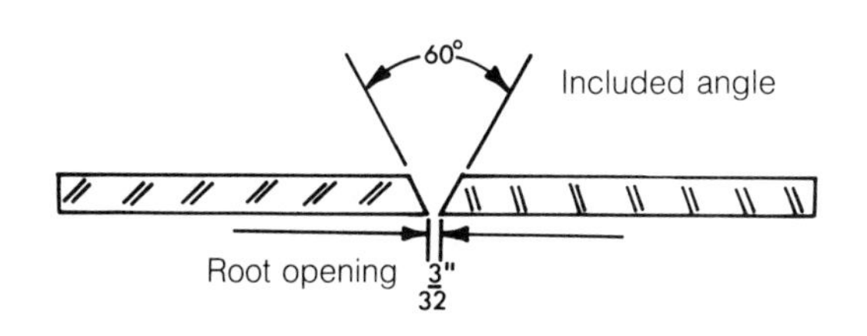


FIGURE 2-33.
Practice welding of heavy plates. Each plate is beveled at a 30° angle, which will make an included angle of 60° for the weld proper.

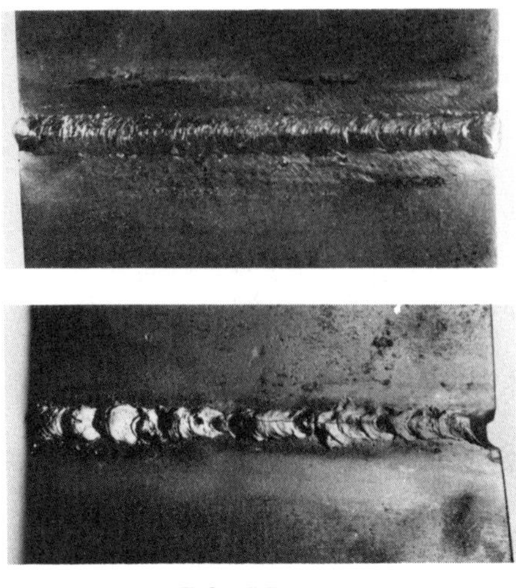

FIGURE 2-34.
Good and bad cover passes. The upper photo shows a properly finished weld, and the lower photo shows an improperly finished one.

***Vertical Position.*** When you can demonstrate that your welds will stand a guided-bend test, you should advance to the vertical position. The vertical position is one of the positions required for certification of acetylene welders.

Time and trouble can be saved in this position by not bothering to practice with 1/8″ or 3/16″ plate. Concentrate on plates of the size used in the certification test. These plates are usually required to be 8″ long, 5″ wide, and 1/4″ thick; however, different states and communities have different thickness requirements. Check with your instructor to be sure. If you take the tests on 1/4″ plate and pass the root and face bend tests, you will be qualified to weld metal twice the thickness of the test plates in those positions regulated by state law and the American Welding Society (AWS).

When practicing the vertical position, you should carefully watch the weld pool to make sure of 100% penetration on the root pass. If you have practiced and paid attention to your instructor, you should have no trouble passing the vertical position test for certification. At the finish of the certification test, two strips 1 1/2″ wide will be cut and ground off flush on both the face and root sides for the 180° bend test. One strip will be bent against the root of the weld and the other against the face. If both plates pass inspection you will be certified for the vertical position only under the limits of the laws in the state in which you reside, as well as under the rules and regulations of the American Welding Society and the American Society of Mechanical Engineers (ASME).

FIGURE 2-35.
A guided-bend test machine is used to make sure all samples are bent under the same conditions.

***Overhead Position.*** The overhead position is the most dangerous and difficult of all the butt joint positions. The best way to weld butt joint plates for certification in the overhead position is backhanded. To get the hang of it, return to ⅛″ plate and weld a square butt joint in the flat or downhand position. But this time, begin on the wrong end of the plate and weld backwards. The rod must be fed over the top of the weld into the weld pool as shown in Figure 2-37.

At first it will seem awkward to weld this way, but it becomes easier after a few tries. The reason for using the backhand method in overhead welding is to ensure positive penetration at the root of the weld.

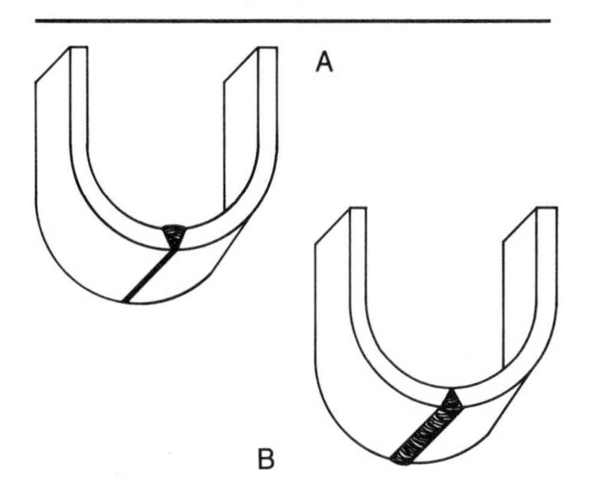


FIGURE 2-36.
Samples are bent 180° across the root (A) or across the face (B).

Before returning to the overhead position, the student should weld 3⁄16″ and ¼″ beveled plates backhand in the downhand (flat) position, from left to right. After you have learned to achieve good penetration in the downhand position, you should practice in the overhead position, using the backhand method. After four or five plates are done this way, the 1½″ strips may be cut, ground, and bent in the guided-bend test to check for penetration and fusion. When the samples can be bent consistently without showing signs of fracture you will be ready for certification tests in the overhead position.

***Horizontal Position.*** In this position, there are two definite torch angles to be concerned with, as shown in Figure 2-38. First, the torch must be tilted upward at about a 70–80 degree angle. Second, it should be angled about 45 degrees in the direction of travel. Because the face of the plate is in a vertical position, the force of the burning gases will aid in holding the molten metal in position as the weld progresses. These angles are rather critical, and you should follow

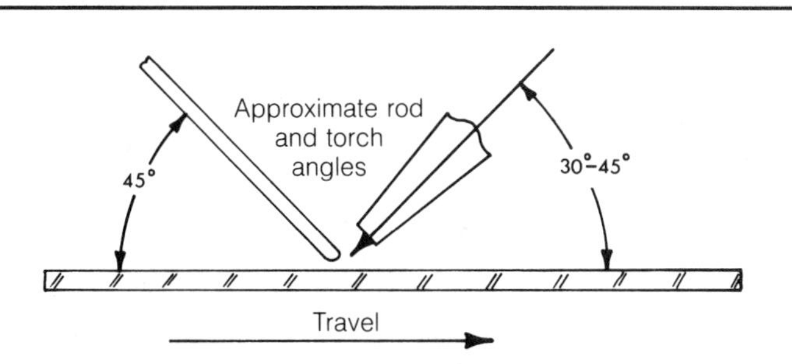


FIGURE 2-37.
The proper rod, torch, and travel direction for backhand welding. This applies to right-handed welders only. Left-handed welders must reverse this process.

them closely in order to obtain a good weld.

In the horizontal position, the way you feed the filler rod is also important. The rod should be fed into the weld at the top of the bead. This will help prevent undercutting at the top of the weld. Never use more filler rod than is needed to make a full bead, as an excess of filler rod will let the molten metal run down and away from where it is needed.

Again, keep the bottom part of the weld bead slightly ahead of the top. This will form a shelf of molten metal on an angle, which makes it easier to keep the weld pool in position.

Start with 1/8" thick material and try thinner and thicker metals after you can easily weld 1/8" metal. Plates of mild steel about 3" × 6" should be used for this exercise. Leave a root opening of 3/32" between the two plates to allow for penetration. Tack weld each end.

After the first tack has been applied and cooled, straighten the plates to be sure they are even. The root opening should remain at 3/32". When a tack weld cools, it pulls the two plates together and causes the root opening at the opposite end of the plate to decrease. Spread the plates to the desired root opening with a screwdriver or some other tool before tacking the opposite end. If the plates were left as they were after the first tack had cooled, complete penetration would be very difficult, or impossible, to obtain.

As in the preceding exercise in butt welding, the keyhole is still necessary to get complete penetration. If this keyhole is not kept constant, penetration will not be obtained. To make this exercise easier, study Fig-

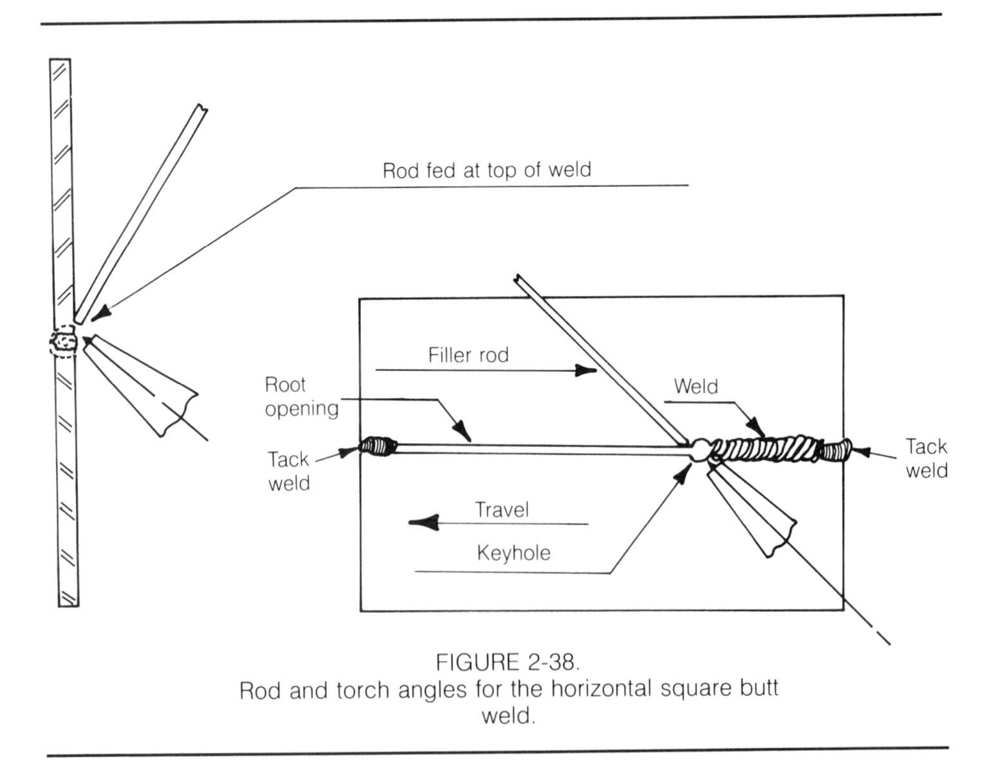


FIGURE 2-38.
Rod and torch angles for the horizontal square butt weld.

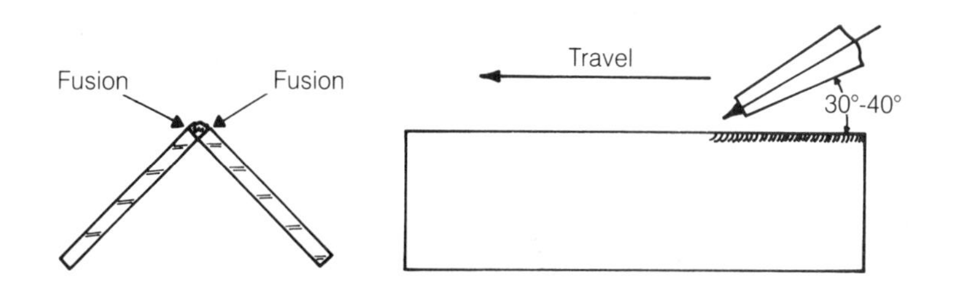


FIGURE 2-39.
Fusion welding. A weld in which no filler rod is used, but the thin edges of the plates are melted to form the weld.

ure 2-38 for the proper rod and torch angles. These angles may vary slightly as the weld progresses.

Note that three of the five welding joints you have made are not practiced before taking the certification tests. These three joints have no direct bearing on the tests and there is no certification for them. The Tee joint is used to teach control of the torch and filler rod, and the square butt joint is the actual certification test plate.

The other joints may be practiced after certification is acquired. For example, a student should learn to carry a 1/4″ bead in a single pass because 6″ pipe is frequently welded on the job in a single pass. For certification, two passes may be used. After the tests are passed, learning these joints can be a pleasant pastime in order to become more proficient. None are difficult. All can be done in all positions. Any thickness of metal can be used, but thinner metals are more logical since welding them requires less heat.

## FUSION WELDING

Fusion welding should be practiced on the corner joint. Fusion welding means melting the edges of the two plates together without using filler rod, as shown in Figure 2-39.

Rapid side-to-side motion of the torch is necessary in fusion welding to prevent the plates from burning through. Tanks, boxes, and many intricate parts are manufactured with this weld. Fusion welding, if done properly, comes out very smooth, does not require finish grinding, and gives you a real opportunity to show off your skill. Large machinery guards are often welded this way.

## LAP JOINTS

The lap joint can be difficult in the vertical and overhead positions. It's bulky and not used too often unless it's really needed. The big problem with the lap joint is burning away the edge of the top plate as shown in Figure 2-40.

The filler rod and torch must be handled differently than on butt or Tee joints to prevent fluffing away the thin edge. Most of the heat must be directed at the bottom plate. The operator must watch closely to be sure the edge of the top plate is melted only enough for good penetration and fusion of the filler rod. The rod is also fed differently. Feed it at the edge of the top plate so that the cold rod will help prevent the thin edge from burning away.

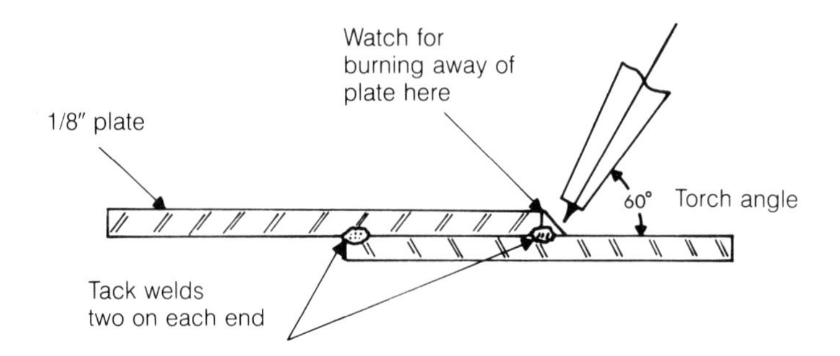


FIGURE 2-40.
Approximate torch angles for the lap joint in the flat position. Careful feeding of the filler rod and expert handling of the torch is needed to avoid burning away the edge of the top plate.

Once again, there are no set torch angles, as they must continuously change as the weld progresses. This joint is difficult in the vertical and overhead position, because heat rises and preheats the plates farther ahead of the weld zone than it does in the downhand position. Also, in the vertical and overhead positions, the operator does not have good control of the torch and rod.

# OXY-ACETYLENE CUTTING

3

Cutting is nothing more then the rapid oxidation (burning) of iron. It is the opposite of welding, even though it uses the same gases (acetylene and oxygen). The torches used for cutting are also different.

## THE CUTTING TORCH

The tip of a cutting torch has 4 or 6 small preheating orifices equally spaced around a large central cutting orifice. Note that in Figure 3-1 the cutting orifice is much larger than the preheating orifices around it. Pure oxygen passes through this cutting orifice, under pressure, and combines with the preheated iron to make instant rust, which is then carried away by the pressure of the gases and gravity. The smaller preheating holes burn a mixture of oxygen and acetylene just as a welding torch does.

Apart from the flame adjustment valves, a cutting torch has still another valve. Usually it's in the form

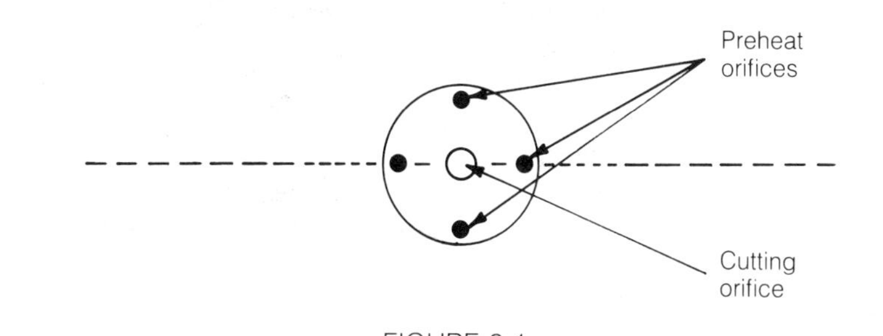


FIGURE 3-1.
Preheating and cutting orifices (holes) of an oxy-acetylene cutting torch.

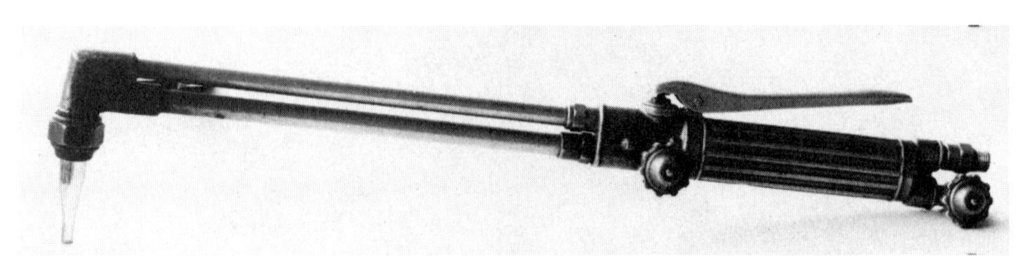

FIGURE 3-2.
A cutting torch. The mixing chamber is located in the torch head. The cutting oxygen valve is the large lever on the handle.

of a small lever located on the top of the torch that is squeezed when the metal to be cut has been preheated to a cherry red.

Most torches used today are combination torches of the equal pressure type. They are economical because their tips can be changed and they can be used for both welding and cutting. When changing these tips, do not use a wrench. HAND TIGHT IS TIGHT ENOUGH. If the torch leaks there's something wrong and it should be repaired.

***Safety.*** The safety rules for welding also apply to cutting, *only more so.* Cutting throws sparks and hot metal much farther than welding. **Make sure all flammable material has been removed from the cutting area. A fire guard must be placed to stop fires from starting.** A fire guard is a person who will stand watch over the surrounding area while you are cutting. *This is especially important where repair work is in progress and all flammable materials cannot be removed.*

***Lighting the Cutting Torch.*** Before lighting the cutting torch, be sure the regulators are adjusted to the proper pressure. What this "proper pressure" is cannot be stated exactly, since it varies for each job, but for openers try using the pressure shown in Figure 3-3. If you're using a *combination torch,* open the oxygen valve on the torch body all the way, since the oxygen for both preheating and cutting must pass through this valve. If you're using a cutting torch, the preheating flame is adjusted by a small valve on the body.

Light the cutting torch just as you would a welding torch by opening the acetylene valve on the body about 1/4 turn. This will let enough gas through the tip to eliminate the soot and smoke that comes with a raw acetylene flame. Continue opening the acetylene valve until the flame starts to jump away from the tip (Figure 3-4A). At this point you're using the maximum amount of acetylene the tip was designed for. Close the acetylene valve until the flame has settled down and is burning at the very end of the tip (Figure 3-4B). The flame will be nearly white and free of any soot or smoke.

When this is done, open the cutting torch oxygen valve and adjust the preheat orifices to neutral flame positions, where the feather disappears and the flames are bright blue (Figure 3-4C). This neutral flame has a

| Thickness of Steel | 1/4″ | 3/8″ | 1/2″ | 3/4″ | 1″ | 1/4″ | 1-1/2″ | 2″ | 2-1/2″ | 3″ | 4″ | 5″ | 6″ |
|---|---|---|---|---|---|---|---|---|---|---|---|---|---|
| Tip Size | 0 | 1 | 1 | 2 | 2 | 2 | 3 | 3 | 4 | 5 | 5 | 6 | 6 |
| Gauge Pressure Oxygen (psi) | 30 | 30 | 40 | 40 | 50 | 60 | 45 | 50 | 50 | 45 | 60 | 50 | 55 |
| Gauge Pressure Acetylene (psi) | 3 | 3 | 3 | 3 | 3 | 3 | 3 | 3 | 3 | 4 | 4 | 5 | 5 |
| Speed in Inches per Minute | 29 | 19 | 17 | 15 | 14 | 13 | 12 | 10 | 9 | 8 | 7 | 6 | 5 |
| Approximate Width of Kerf in Inches | .075 | .095 | .095 | .110 | .110 | .110 | .130 | .130 | .145 | .165 | .165 | .190 | .190 |
| Cutting Orifice Cleaning Drill Size | 64 | 57 | 57 | 55 | 55 | 55 | 53 | 53 | 50 | 47 | 47 | 42 | 42 |
| Preheat Orifice Cleaning Drill | 71 | 69 | 69 | 68 | 68 | 68 | 66 | 66 | 65 | 63 | 63 | 61 | 61 |

FIGURE 3-3.
Gas pressures used for different tip sizes and metal thickness.

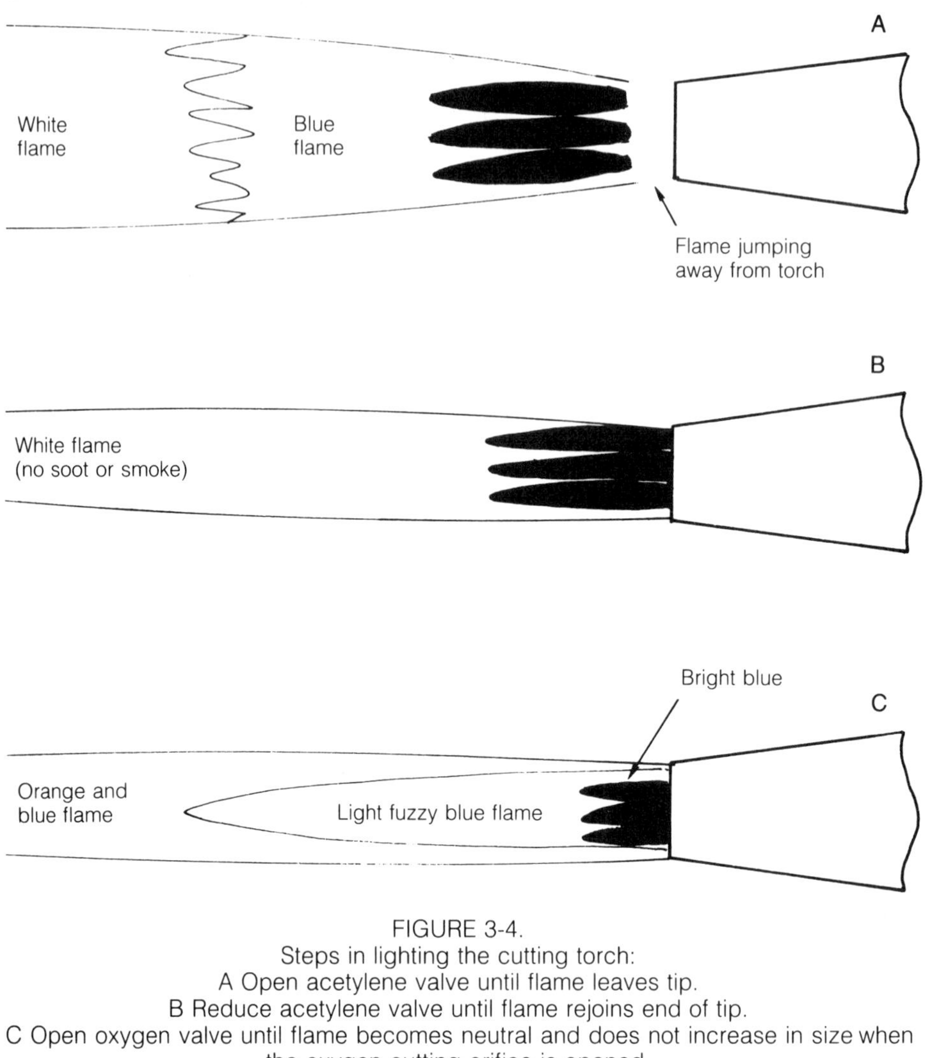


FIGURE 3-4.
Steps in lighting the cutting torch:
A Open acetylene valve until flame leaves tip.
B Reduce acetylene valve until flame rejoins end of tip.
C Open oxygen valve until flame becomes neutral and does not increase in size when the oxygen cutting orifice is opened.

temperature of about 6000°F. When the flame has reached this temperature, you're ready to cut. The proper name for the slot that the torch will cut in the steel plate is a "kerf."

## CUTTING

Place the torch at the edge of the plate you're going to cut, letting about half the flame spill down over the edge of the plate. When the edge of the plate turns red and starts to look wet, squeeze the oxygen cutting valve lever, keeping the tips of the preheat flames 1⁄16″ to 1⁄8″ away from the surface of the plate, and the sparks will fly as the cut begins.

If you let the preheat flames dip into the molten metal at the top of the plate you may ruin the tip. Cutting tips are made of nearly pure cop-

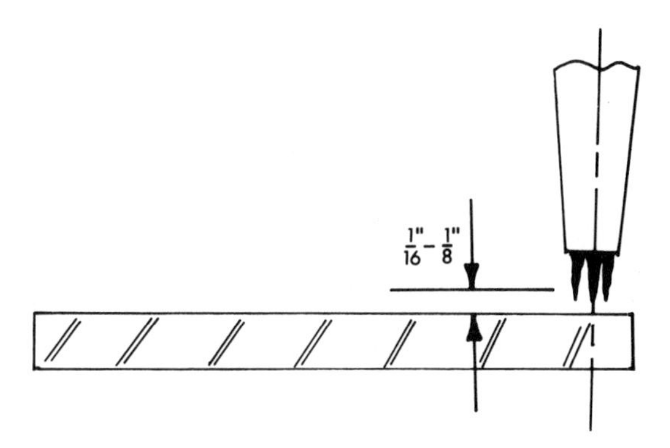


FIGURE 3-5.
Diagram showing the proper distance to hold preheat flames from the surface of a plate while making a cut.

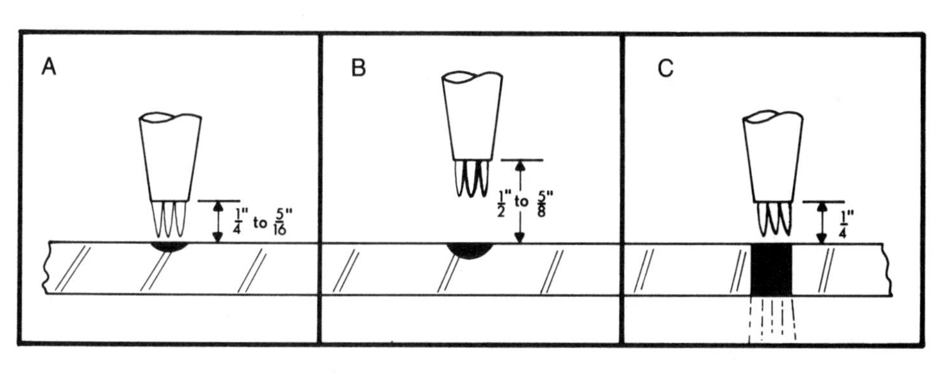


FIGURE 3-6.
Procedure for cutting holes in steel plate.

per and they melt at a much lower temperature than steel.

Try to keep the speed of the torch across the plate even, traveling as fast as you can without stopping the cut. If you move the torch too fast the preheat flames will not have time to heat the metal to its kindling point, 1600°F, and the cut will stop. Excessive speed can also force molten metal out of the kerf onto the top of the plate, which again endangers the life of the tip. If the cut is too slow, the metal being blown from the underside of the kerf will tend to weld itself back together.

**Don't use the torch head as a hammer to knock the cut plate loose.** If the torch head breaks where it fastens to the main body, while recovering from burns, you can console yourself with the knowledge that you'll never work in that shop again. Cutting looks simple but it requires a steady hand which, in turn, requires lots of practice.

If the speed of the torch is right, the pressures correct, and the torch held and adjusted properly, the kerf will be smooth, and very little slag will show on the underside of the plate.

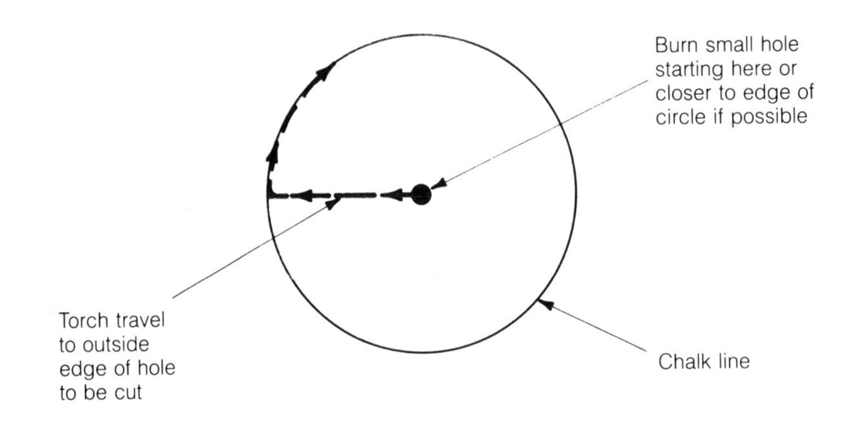


FIGURE 3-7.
Procedure for cutting a large hole in steel plate. The small beginning hole may be started about 1″ from the outside circle to be cut.

***Cutting Holes.*** To cut a small hole, hold the tip of the torch about ¼″ away from the surface where the hole is to be cut (Figure 3-6A). After heating a small area, raise the torch slightly to enlarge the heated area (Figure 3-6B). Lower the torch and depress the cutting lever (Figure 3-6C). To cut a large hole, outline the area to be cut with a piece of chalk. Burn a small hole in the center, and gradually work to and around the chalk line (Figure 3-7). If the hole is large enough you can use a cutting compass like the one shown in Figure 3-8.

***Cutting Round-Bar Stock.*** Hold the torch vertically and slowly follow the outline of the bar. Maintain this

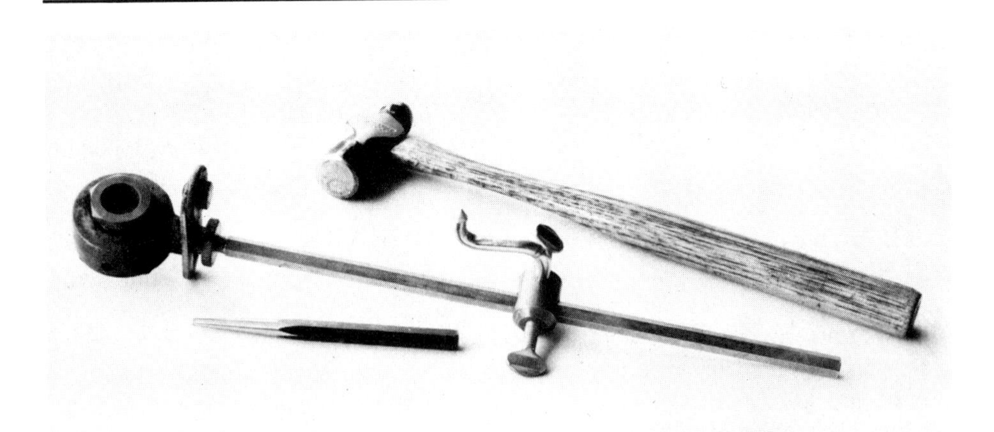

FIGURE 3-8.
A cutting compass.

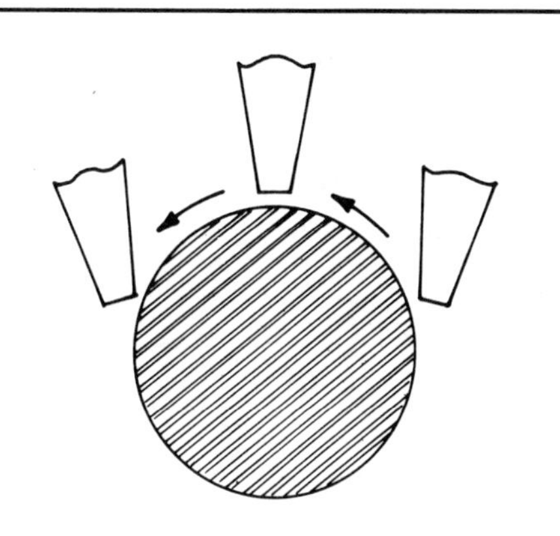

FIGURE 3-9.
Procedure for cutting round bar stock. A small nick may be made in the bar with a chisel or center punch to aid in preheating the area to be cut.

vertical position while ascending as well as descending on the opposite side (see Figure 3-9).

***Automatic Cutting Machines.*** A common method of cutting is with a straight line rig. The torch on a straight line cutter usually has larger tips and can be accurately adjusted for travel and speed, which enables it to cut through much thicker steel than a hand cutting torch. For straight line cutting these machines run on rails. For circle cutting they are usually driven by the same electric motor, but are locked onto a trammel bar. Cuts up to 48″ deep can be made with these power-driven torches.

Cutting steel with the oxy-acetylene process is nothing more than a rapid oxidation process. The rust that appears on steel left in the damp is chemically identical to the slag left under a steel plate cut with an acetylene torch.

Occasionally, propane or MAPP gas are substituted for acetylene when oxy-cutting ferrous metals. Propane is cheap. Acetylene is hotter than propane or MAPP (methyacetylene propadiene), but MAPP is more stable. MAPP can be piped at higher pressures than acetylene, and thus is often used for underwater cutting. Acetylene's 15 psi maximum safe pressure can't push gas out of the torch at depths below 37 feet.

OxyMAPP cutting tips resemble oxy-acetylene tips but hole sizes are greater. OxyMAPP flame cones are twice as long as oxy-acetylene. Cutting with oxy-acetylene, the preheat flame cones touch the metal. Using MAPP, back the torch off to the "coupling distance." The double length flame cone and the coupling distance puts a MAPP torch farther from the metal than an acetylene torch. The inner cone of an OxyMAPP flame must never touch the puddle. If you have to MAPP weld with acetylene tips, go one size larger.

OxyMAPP will weld aluminum, but only in an emergency. Nearly any other process is better.

## NONFERROUS METALS

Most metals cannot be cut with a cutting torch; for example, nonferrous metals such as aluminum, brass, and copper. Some stainless steels can be cut with acetylene, but most cannot. There is a rig for cutting stainless steels which uses a tip rather like the oxy-acetylene tip, but, in addition to preheating orifices and a cutting oxygen orifice, it has additional arrangements to blow a spray of fine iron powder into the cut. This iron powder burns with the oxygen and cuts away the stainless steel, which normally has too much copper and nickel to burn with an oxy-acetylene torch.

There are other attachments for the cutting torch which permit copying from a pattern or template, or cutting several plates at one time. There are also pantograph attachments which permit scaling up or down from a pattern. The patterns can be simple templates or, in more modern devices, they can be magnetic or electronic traces. These are production processes and do not much concern the welder who is using the hand torch for cutting.

# 4 MISCELLANEOUS PROCESSES

## BRAZING

Brazing is not really a welding process because the metals being joined together are not melted. Like soldering, it is an adhesion process in which the metals being joined are heated only to the somewhat lower melting point of the metal used to join them (see Figure 4-1). Brazing can describe several processes, but usually means the joining of iron or steel with brass filler rod.

Brass rod melts at just above 800°F, at which point iron is only red hot. No weld can be stronger than the metals involved in the process; therefore soldered joints can never be stronger than the relatively low tensile strength of a lead-tin mixture. Brass, which is a mixture of copper and zinc, has a much higher tensile strength, and a brazed joint done with care can be *nearly* as strong as a true weld.

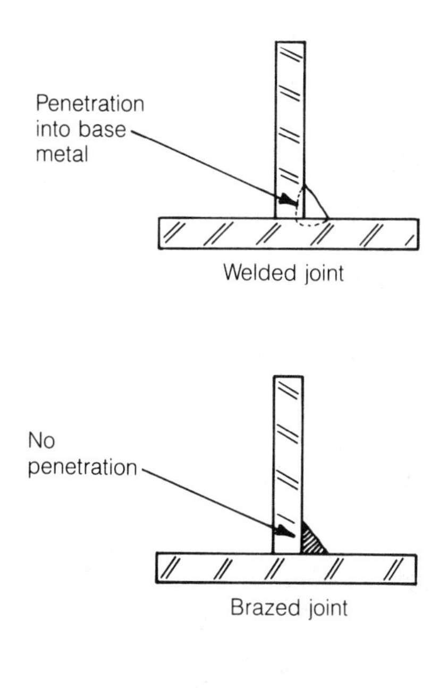


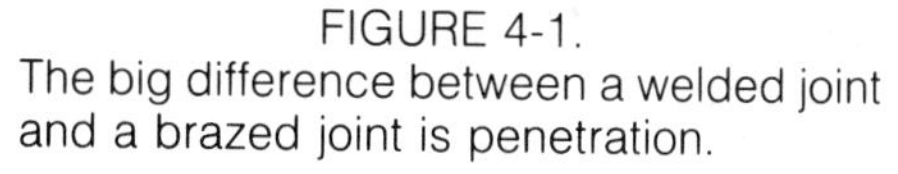
FIGURE 4-1.
The big difference between a welded joint and a brazed joint is penetration.

FIGURE 4-2.
The easiest way to dispense flux is through a hole punched in the top of the can.

***Fluxes.*** Flux is a wetting agent used to clean surfaces to be joined and to exclude air from the weld zone (Figure 4-2). There are many kinds and trade names, but the main ingredient in any brazing flux is borax. Technically, it's sodium tetraborate ($NA_2B_4O_7 \cdot 10H_2O$), but practically speaking, it's "20 Mule Team," and a lot cheaper in the supermarket than in a welding supply house. Flux may be a paste, a powder, or it may come on the brazing rod.

***Brazing Rods.*** Brazing rods come in various alloys, but the most popular is a combination of copper and zinc. The fumes from these rods are poisonous. **Try *never* to breathe fumes from brazing!** Zinc poisoning will make a welder very ill, and there is no cure for the damage that white smoke does to one's lungs.

Other brazing alloys may be made of nickel and chrome, copper and phosphorus, or alloys of silver. Sometimes brazing alloys are even made of aluminum and magnesium. These alloys allow dissimilar metals to be joined (see Figure 4-3).

## BRAZING POSITIONS AND PROCEDURES

The joints in brazing are the same as in any other welding process. For maintenance work, where almost anything can happen, they may not be perfect Tee or groove joints. Lap joints are popular in brazing because the filler metal will flow into thin

Brazing and Braze Welding Alloys

| Alloy | AWS-ASTM* Classification |
|---|---|
| 1. Magnesium Alloys | B Mg |
| 2. Copper and Gold Alloys | B Cu Au |
| 3. Nickel and Chrome Alloys | B Ni Cr |
| 4. Copper and Phosphorus | B Cu P |
| 5. Silver Alloys | B Ag |
| 6. Aluminum and Silicon Alloys | B Al Si |
| 7. Copper and Zinc Alloys (Brass) | B Cu Zn |
| 8. Copper | B Cu |

FIGURE 4-3
*American Welding Society-American Society for Testing and Materials

openings and provide a wide and very strong bond to the base metal.

Brazing is a simple process. The acetylene torch is most commonly used, but brazing may also be done with mixtures of natural gas and oxygen, or propane and oxygen. The amount of heat necessary will vary with the type of metal being brazed. This is something only experience can teach.

For a first exercise try a common Tee joint. Adjust the torch to a neutral flame so that the strength of the brazed joint will not be lessened. A reducing flame (with an excess of fuel gas) will make a very smooth and neatly brazed joint, but a reducing flame will also greatly reduce the strength of the weld.

After tacking, heat the base metals very carefully to red hot and begin applying fluxed brazing rod to the joint. The brazing rod may come already coated, or may be dipped into a wet flux paste, but usually an inch or so of the rod is passed through the torch flame momentarily, then dipped into dry powdered borax. Enough borax sticks to the hot rod to flux the joint. Borax, when heated, glows brilliantly so a darker goggle lens is advisable.

Torch angles for brazing are about the same as those used for welding the same joint with steel welding rod. The big difference is in the torch motion, which should be slightly faster to keep the pool of braze agitated. Braze metal will flow into the joint better if agitation is maintained. Feed the brazing rod steadily and the joint will appear smooth, with even ripples, and with a slight line of penetration on the opposite side of the joint. Brazing is not hard to learn.

At one time cast iron brazing was a popular way to repair broken castings, which were expensive to replace. Cast iron brazing is much more difficult than mild steel brazing and cannot be done at all unless the iron has been totally cleaned of oil and grease. This was usually done with acid baths. Today cast iron brazing has been replaced by arc welding with special nickel alloy electrodes. These special alloy electrodes make a stronger weld and are easier to apply.

## WELDING STAINLESS STEEL

Stainless steels are also brazed, but never with brass rod, because brass rod makes stainless steel crack. Stainless steels may be brazed with a silver alloy rod more easily than welded. This silver alloy rod is the same thing as silver solder that is used in making fine jewelry. It usually contains 45% silver, and it isn't cheap. Silver braze rod makes very smooth joints and does not change the color of stainless steel. Unfortunately, the joints are only good for about a year.

Stainless steel is a very smooth, dense metal with no roughness or porosity for the braze or solder to get a bite into. Brazed joints flow beautifully and test perfectly on any machine, but the elements immediately go to work, and corrosion begins creeping under the edge of that perfect seam.

The elements slowly lift and pry the silver braze up from the stainless until, about a year from the time it was brazed, the joint will come apart. The stainless steel and silver brazed

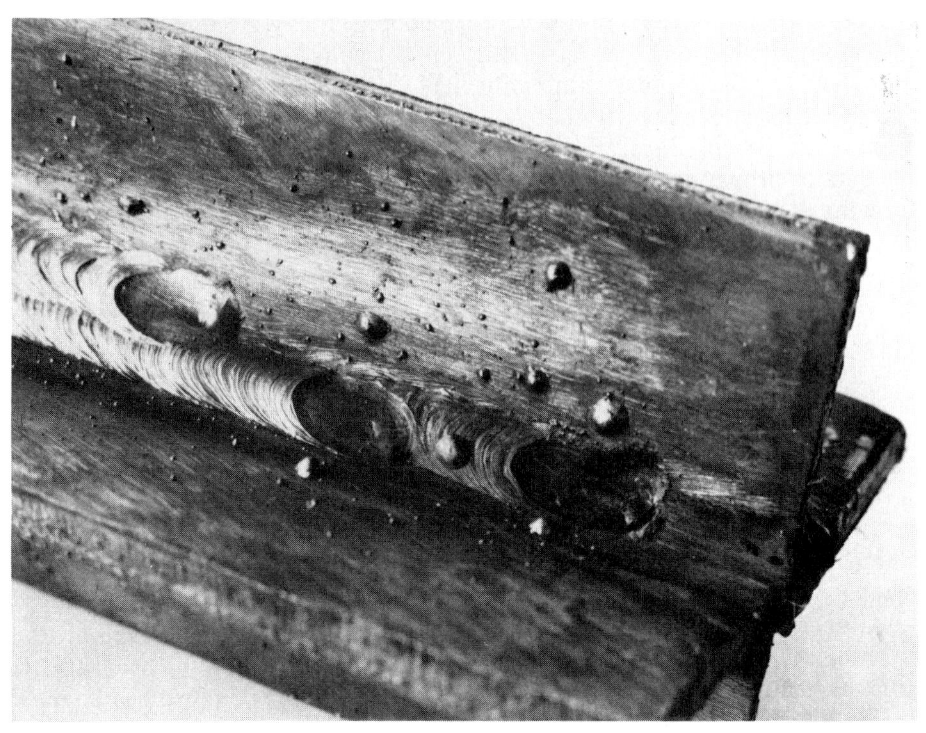

FIGURE 4-4.
The tiny droplets that appear when welding stainless steel are called spatter.

surfaces will still be glassy smooth, but they will no longer be stuck together. It was something like this that kept the nuclear submarine Thresher from surfacing the last time it dived.

There are other, safer processes for joining stainless steel. They may not be quite as pretty as a smooth silver brazed joint, but they hold together longer. Electric arc welding is used on heavier gauge metals which can resist the heat distortion. The electrodes may be used in all positions of welding and with any type of joint, but you'll learn more about this later.

The only objections to arc welding stainless steel are discoloration and spatter (Figure 4-4), which can largely be prevented by painting the work with an anti-spatter compound before welding. If there's no anti-spatter compound available, go to the washroom and snitch some liquid soap.

Stainless steel electrodes cost more than ordinary steel, but they flow so smoothly that they're often used to join plain steel if the job happens to be particularly tricky.

Make sure you have good ventilation. Stainless steel electrodes produce fumes that are very hazardous to your health.

TIG welding is a neater and better way of joining stainless steels. This process, which will be discussed in detail later, is used extensively in the aerospace industry and for manufacturing the stainless steel tanks used in dairies, wineries, and breweries.

On blueprints, stainless steels are usually called CRES, which is confusing because CRS usually means *cold rolled steel*. CRES stands for *corrosion resisting*, which stainless steel certainly is. It's really iron with about 18% chromium and 8% nickel mixed in. Chromium and nickel are *highly* resistant to corrosion, and they give some of their corrosion-resistant properties to the iron. Because of these percentages there's a whole series of stainless alloys which are called 18-8 steels. Some CRES steels are magnetic, but most are not. The high temperatures which must be used to weld stainless steels tend to burn out some of the chromium and nickel, and any stainless steel will usually show a slight rust streak along the weld if used under extreme conditions—like under salt water. Stainless steels are very tough but do not temper well.

## WELDING ALUMINUM

Aluminum can be welded with oxy-acetylene but the process is seldom satisfactory. Oxy-hydrogen works better, but this process has largely been supplanted by TIG and MIG processes, which are faster and create cleaner welds in aluminum. TIG and MIG welding techniques are covered in chapters 7 and 8.

## WELDING COPPER

Copper conducts heat even faster than aluminum and requires very large tips. Copper is also hot-short, which means it tends to collapse when near the melting point, so weldments must be well supported. Copper is usually welded backhand with a neutral or slightly reducing flame and a fluxed rod.

## SOLDERING

Probably the easiest (and oldest) way of joining metals together with the least amount of distortion is with solder. There are many kinds of solder—one actually works on aluminum, but with even less reliability than silver brazing works on stainless. One speaks usually of *soft* solder, as opposed to silver solder, which is actually a braze; soft solder refers to a mixture of lead and tin. The two metals can be mixed in any proportion, but are usually 50-50. Sometimes the proportion of tin is raised to 60-40 to lessen the danger of lead poisoning when fabricating food handling equipment. The percentage of tin can go even higher—to 80-20 or 90-10.

The odd thing about many alloys is their melting temperature. Lead melts at 620°F. Tin melts at 450°F. Mix them together and the melting point is not somewhere between these two figures as might seem logical. Instead, a half and half lead-tin mixture melts at 357°F!

Plumbers use a lot of 50-50 solder. Copper tubing does not rust out as quickly as iron pipe, nor does it take as long to solder a joint as it does to cut and thread pipe.

Because of the low tensile strength of soft solders, butt and Tee joints are seldom successful. A properly prepared lap joint, however, will usually be as strong or stronger than the parent metal.

Fluxes must always be used to clean the metal, since solder will not flow onto dirty metal. These fluxes, depending on the metals and other conditions, can be raw muriatic acid, cut acid, which is the same thing after it's been killed with pieces of zinc; or various commercial fluxes usually guaranteed not to corrode. Electrical work is often done with rosin flux or with rosin cored solder wire.

Heat for soldering can come from any source. Soldering coppers can be heated with gasoline or propane blowtorches, but most tinbenders prefer charcoal since the heavy oil residuals in petroleum-based fuels sometimes make the copper (which for some reason is usually called a soldering *iron*) hard to tin. Soldering irons (coppers) come in several sizes and shapes for many different jobs. Some are heated electrically, and there is one electrically heated iron with a small tip which heats so rapidly it is called a gun, and burns away almost as quickly as it heats up. These *guns* can be useful for quick, one-shot jobs like joining ferrules to the ends of motorcycle control cables.

Acetylene, natural gas—any heat source can be used to heat a soldering iron. The main problem with acetylene is that it's too hot, and too much heat will oxidize the copper and leave it covered with a black scale, which must be wire brushed off. Smaller accumulations of crud can be removed by dipping the hot iron into a solution of sal ammoniac (also called ammonium chloride), or rubbing the iron over a solid block of sal ammoniac. Sal ammoniac sends up a great stink, but isn't all that poisonous. It is even used, in small amounts, in cough syrups.

Soldering is often done without an iron, by using a portable propane torch to heat the metal until solder will flow. This is a great way to solder copper pipe, and plumbers use propane torches almost exclusively (Figure 4-5).

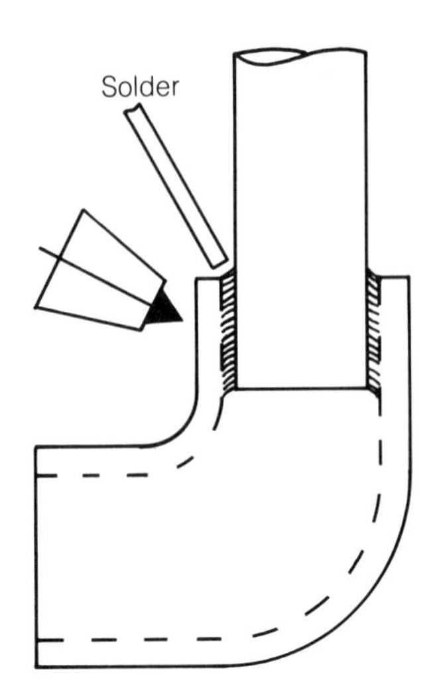


FIGURE 4-5
A propane torch is often used to solder a lap joint on copper tubing.

In soldering, remember that the metals must always be clean. Flux cannot do it all. If the metals are dirty or heavily corroded, then sand, grind, wire brush, scrub with stainless steel wool, or otherwise clean the surfaces to be joined before fluxing. Heat only the metals to be joined; never the solder. Some solders have rosin or acid fluxes in their core, but these are usually effective only when joining shiny new wire or tubing. In all other cases, flux the cleaned metals; heat them to the melting point of the solder, taking great care not to get the metals too hot. Too much heat burns away the flux, dirties the surface of the metal, and, in general, creates far more problems than trying to work with metals that are too cold.

A properly prepared soldered joint is truly marvelous to behold. When the metals to be joined are cleaned, fluxed and heated to just the right temperature, there seems to be an instantaneous little happening as the operator touches the solder to the joint, and the solder suddenly flows everywhere, creating a perfect joint in a fraction of a second. But it takes perfect cleanliness and heat control, and these take practice to achieve.

Any soldering operation should be supported in some way. This can be done with C clamps (Figure 4-6), jigs, or other devices for holding the pieces steady. If the work is jiggled while the solder is cooling, the joint will be ruined. Any soldering should be done as fast as possible, and cleaned of the excess solder and flux immediately.

Semi-automatic processes are sometimes used to solder when many similar parts are being produced. This was very common in the food canning industry until modern machines were developed which could seal cans without solder.

The real skill in any soldering process is in the preparation of the surfaces to be joined. If the surfaces have been properly cleaned and fluxed, the solder will flow.

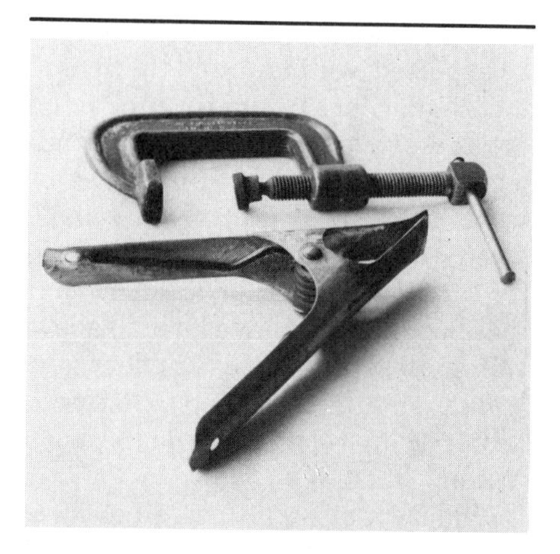

FIGURE 4-6
Two types of clamps that may be used to hold your work steady when welding or soldering.

## HARDFACING

Hardfacing is the process of fusing a very hard and wear-resistant metal to a surface that will be subjected to extreme wear or abrasion. Sooner or later, any combination welder will have to do it. This is usually a maintenance project, on things like the edges of bulldozer blades, but hardfacing can also be done to reduce the wear on parts to be used under unusually severe conditions.

Hardfacing can be done several ways. One method is with the acetylene torch, which is easier to control than the extreme heat of an electric arc. The overheating of an electric arc may also cause poor bonding between the hardface and base metal.

Hardfacing is used to build up parts subject to high friction and rapid wear, such as caterpillar treads or the blades of earth moving equipment. The type of hardfacing rod used will depend on the alloy of steel that is to be covered. There are many different types, and each has its purpose. A popular one is called tube borium. This rod is used with acetylene and consists of a thin shell of mild steel *lightly coated* with copper. The inside of the rod is filled with borium powder.

A carburizing (reducing) flame is used to apply this rod. This reduces the amount of heat applied to the weld zone and makes the application much easier and of higher quality. A neutral flame will heat the base metal too much and cause cracking along the edges and underneath the weld. Also, the carburizing (reducing) flame mushrooms more and helps preheat the base metal ahead of the weld zone. This helps give the weld a smoother appearance and improves the bond of the rod material to the base metal.

Postheating (playing a torch over the work to prevent rapid cooling) is also important here. Slow cooling increases the quality of the hardface and helps prevent cracking of the base metal.

Hardface may be applied much faster with the electric arc than with acetylene, but, if done incorrectly, the weld will be inferior. Hardface may also be applied with the TIG welder and by another method known as metal spraying. The TIG process is expensive but produces a very high quality weld. Metal spray is used extensively in rebuilding crankshafts for auto repair shops. Metal spraying is done with a special acetylene torch where the hardface may be applied from a container in powder form or via a specially built wire feeder. The hardface is melted in the acetylene flame and actually blown onto the work surface. We haven't studied those processes yet, but you'll get enough arc and TIG welding later on.

In any of the acetylene processes, the tip used for hardfacing will be about two sizes larger than would be used for welding with mild steel rod. This larger tip is necessary for the preheating that must be done ahead of the work.

## WELDING BRONZE

Bronze welding is done mostly with acetylene; other methods are TIG and electric arc with shielded metal electrodes, but these two processes are rather expensive and are not used as much as acetylene. Bronze is an alloy of copper and tin. Neither metal will change much in color when melted, and many welders get a bad weld because it is hard to guess its temperature from the bronze's appearance.

When welding bronze, use an oxidizing flame to eliminate toxic fumes. Bronze fumes are poisonous, but not nearly as bad as those of brass.

If plates are to be welded with a butt joint, they should be prepared before welding by chamfering or beveling and by using a backup plate. When arc welding bronze, the electrodes (rods) are usually heavily coated. The amperage used for welding bronze must be much higher than when welding mild steel. The electrodes should also be of the same alloy as the metal being welded, but we'll get into that in the next chapter.

## UNDERWATER WELDING

Metals can be oxy-acetylene or arc welded underwater. The work is dangerous and opportunities for electrocution abound. Special training, special equipment, and many safety precautions are required. As might be expected, water around the heated metal quenches it rapidly, producing hardness (martensite) in the heat-affected zone. Therefore, underwater welds have about 80% of the tensile strength and only 50% of the ductility (ability to bend or stretch) that these welds would have if made above water. Because of this, underwater welds should be proportionately larger than those performed above water.

Underwater welds are seldom permanent, and are usually intended as temporary repairs until the structure can be taken out of the water.

# 5 PRINCIPLES OF ARC WELDING

## ARC-WELDING MACHINES

There are many types and makes of arc welding machines, each of which has its purpose, even if only to serve as a horrible example. The purpose of this book is not to advertise machines. The choice of which machine to use depends on the kind of work to be performed and the amount of money a buyer is willing to spend.

Arc welding differs from acetylene welding in that the arc welding machine uses electricity instead of gas as a heat source. Sometimes arc welding is done with a carbon electrode, while filler rod is fed separately, as in acetylene welding, but usually the coated filler rod also serves as the electrode.

AC (alternating current) welders (called sputter boxes by the pros) are cheap and a favorite of the home welder, but they're for light duty

FIGURE 5-1.
An AC arc welder. Turning the crank clockwise will increase the current.

FIGURE 5-2.
An AC-DC rectifier welder that will produce three types of current: AC, DC (straight polarity), and DC (reverse polarity).

FIGURE 5-3.
A DC-generator welding machine. This machine is driven by an AC motor attached directly to a DC generator.

FIGURE 5-4.
A DC-generator welding machine that is driven by a gasoline engine. The gasoline engine is connected directly to a welding generator.

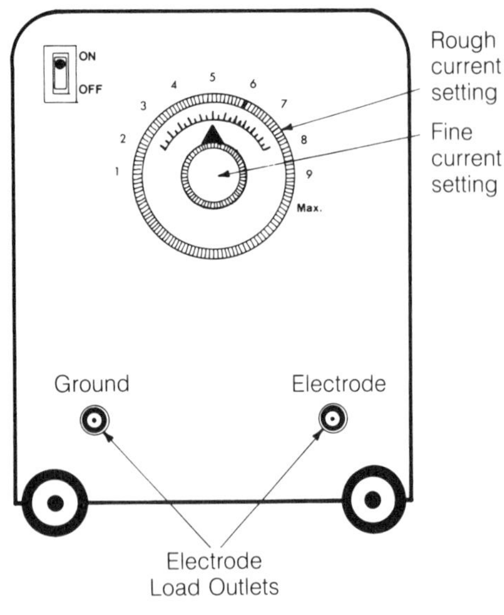

FIGURE 5-5.
An AC welding machine. Current for welding is raised or lowered by the hand wheels on the face of the machine.

only. There are also more expensive, heavy-duty rigs which rectify AC into DC. The third and most common type, however, consists of an electric AC motor (or sometimes a gas engine) which drives a direct current generator.

The theory of electricity, according to which expert you believe, has current flowing from negative to positive or positive to negative. For some welding it doesn't make any difference. Sometimes the welder will hook up polarity one way, sometimes another, depending on the metal being welded and the kind of rod.

It might be helpful here to review Ohm's Law, which states, among other things, that voltage and amperage are dependent on each other. Volts times amps equals watts. 1 volt times 100 amps equals 100 watts. 2 volts times 50 amps equals 100 watts. 100 volts times 1 amp and you've still got 100 watts.

Basically, what any welding machine powered from an AC line does is to change high voltage, low amperage current into low voltage, high amperage current. Amperage kills but voltage gives it the punch. It's pretty hard to get a shock from a welding rig, but you can get burnt in a hurry. **Watch it!**

As in any other kind of welding, the thicker the metal, the more heat required. In arc welding, more heat means more amperage, and that's why arc welding machines have controls to set voltage and amperage according to the work being done. Some of the fancy ones even have foot pedals to give the work an extra hot shot at the start of a cold plate, and back off on the power near the end of the weld where the heat has nowhere to go and suddenly builds up to cause a blowout.

### AC Welder

First, let's consider the AC welder. It has other, fancier names, but the first time you use one you'll know why it's called a sputter box. Most AC welders are cheap, and many small shops use them for this reason. A DC unit capable of the same amperage costs three times as much. One nice thing about sputter boxes is a lack of "arc blow." Arc blow is what happens when a current of electricity wavers from its normal path of travel. It's caused by magnetic fields that are distorted by the shape of the work being welded, and soon you'll know more than you want to know about it.

### AC-DC Rectifier Welder

The AC-DC rectifier welder has no moving parts except cooling fans. They're not very portable, but they can be used for every type of welding—SMAW, TIG, MIG, straight (DCSP), or reverse (DCRP) polarity, and they usually offer greater flexibility in voltage and amperage than motor-generator type machines.

There are many makes of welders, and they all work. Some are better than others, some are best for one purpose, and some are best for other purposes. This could be the subject for a book in itself, but since such a book would also be the subject of countless lawsuits, don't hold your breath waiting for this author to write it.

## ARC-WELDING EQUIPMENT

Apart from the welding rig, stinger, ground clamp, and cables, one needs proper clothing. With acetylene welding it's sometimes possible to scrape by without everything, but arc

FIGURE 5-6.
A portable DC welding machine with the equipment needed for welding.

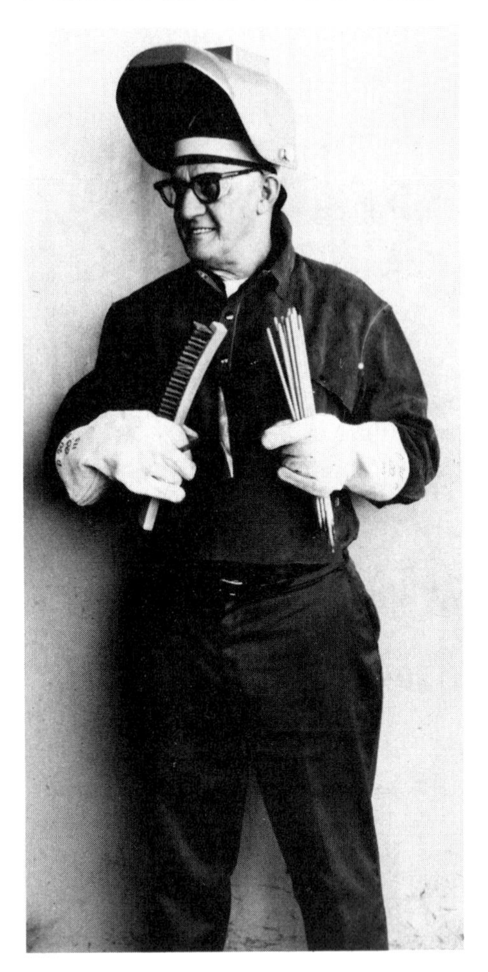

FIGURE 5-7.
A welder properly dressed for electric arc welding. Notice the safety glasses (a must for chipping slag), and the leather sleeves and cape for protection from ultraviolet rays generated by the electric arc.

welding *requires* a full outfit of at least two layers of clothing. Wear a "T" shirt and a long sleeved shirt, preferably cotton of a dark color.

Welders should *never* wear synthetic materials of any kind. Cotton and wool—natural materials—will burn. Synthetics *melt.* A hot, sticky gob of melted nylon or other synthetic material often cannot be torn loose. It will char human flesh much deeper than the quick flash of flame that melted it.

The electric arc welder emits ultraviolet rays just like the sun does. The difference between the arc welder and the sun is that the arc is roughly 93 million miles closer! Welding without proper clothing means severe burns at best. Do it more than once or twice and it means skin cancer. Cover up all those little places people forget, like the open V neck of a shirt and the gap between your shirt cuff and glove. You won't notice it when welding, but later that night you'll learn that sunburn, by comparison, is positively refreshing.

Scientists have come up with some pretty amazing miniature radar sets,

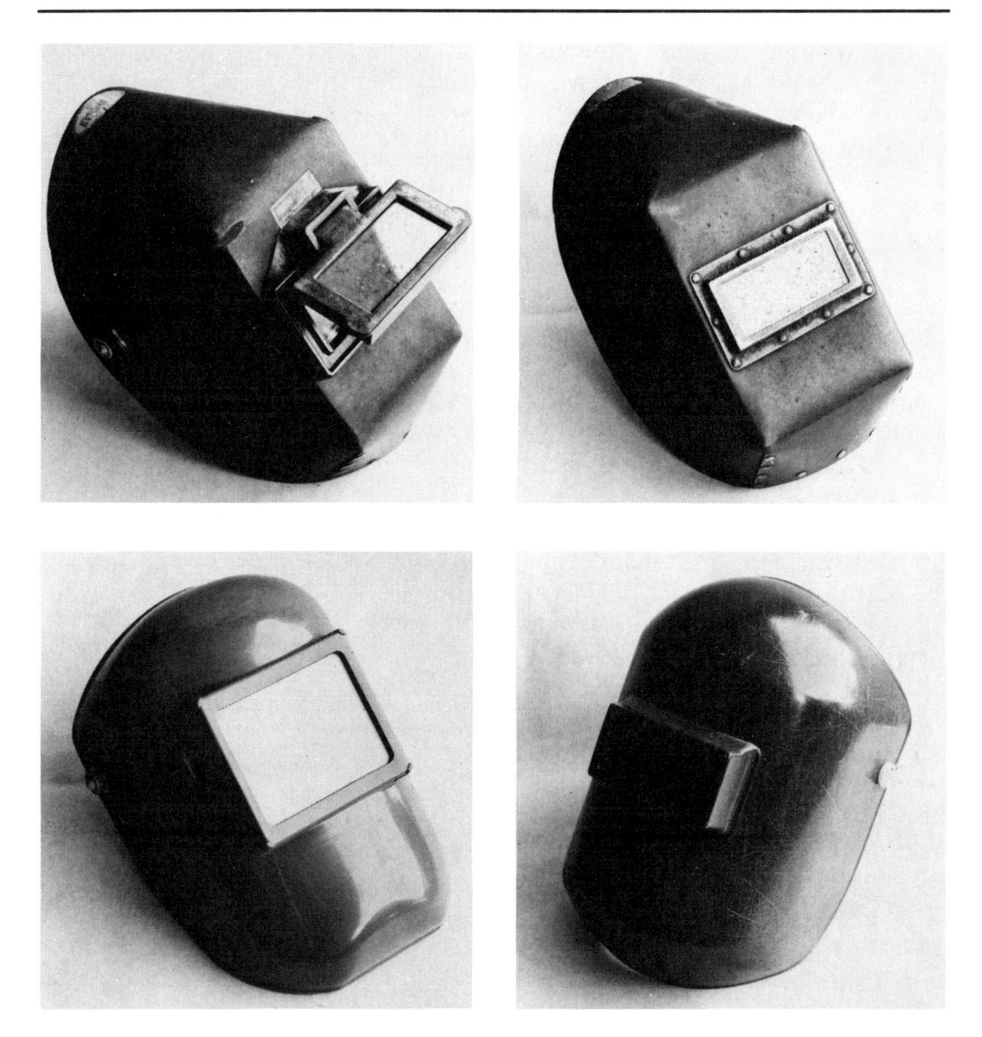

FIGURES 5-8—5-11.
Different styles of welding helmets. There are many styles of welding helmets—some are lighter than others, some have wider fields of vision, and some are designed to be worn over glasses.

but there are still no reliable replacements for eyes, and you can burn out your only pair much faster than your skin. It takes less than a second for a "flash" to come when you least expect it—often from raising your hood just as the welder next to you is striking an arc. It's even more humiliating to cause your own flash burn by carelessly putting down a stinger where the live electrode can ground. At the moment of "flash" you will experience nothing more than annoyance and a few moments of blindness, but when you wake up at midnight with your eyes full of sand and feel more pain than you ever imagined could exist in only two eyes, then you'll know what "flash" is. If you get a flash, see a doctor or nurse at once. They can't prevent all the pain but they can make it better, and *perhaps* they can prevent severe eye damage. People with gray, green, or blue eyes are more susceptible to flash. One number darker glass in hood or helmet can usually compensate for this greater sensitivity.

And so we come to the helmet or hood. These helmets are usually made of black fiber material that looks like heavy cardboard. They keep sparks and radiation from the welder's face. Sometimes they have a little leather apron at the bottom to cover the open V in a shirt. They have a cover glass to catch slag and spatter and behind this is a special lens to filter out about 98% of the ultraviolet radiation from the electric arc. These lenses come in varying shades of darkness. Usually a #10 is about right. As in any other kind of welding, heavy work requires a darker shade and vice versa. Don't be surprised if you can't see anything through this lens. An electric arc is more than twice as hot as an acetylene flame. Even when the arc is struck you can't see beyond the edge of the weld pool.

Gloves should be well insulated and made of leather. The gauntlets should be long enough to come up over your shirt cuffs. They're expensive, so don't ruin them by picking up hot metal.

Welding leathers are not like ordinary clothes. The sleeves come separate from the rest of the outfit. When welding vertical or overhead, no matter how hot the day, no matter how sweaty the welder, full leathers must be worn. The alternative is even more unpleasant. Leather aprons or chaps are also helpful. Pants cuffs catch hot metal and are dangerous. Low shoes seem to draw great gobs of molten metal that would go elsewhere if the welder were wearing boots. New leathers are expensive and many jobs do not require a full set. Students can save money buying second hand

FIGURE 5-12.
A good pair of welding gloves is a dire necessity in the welding trade. They come with various lengths of gauntlets and different types of insulation to prevent burning your hands.

FIGURE 5-13.
Leather sleeves are a necessary part of the welder's attire. They are designed to protect you from ultraviolet rays, molten metal, and sparks.

leathers, since they may never again use them after training.

Welders' shoes are very important. As long as *you're* buying them, invest in a pair of safety boots that have a steel cap inside the toe. This can take the weight of a dropped oxygen bottle much less painfully than your toe. The top of your shoe or boot should be high enough to be covered by your trousers or it will become a funnel for hot iron.

Leather soles are slippery, and they are usually put on with iron nails, which are a pretty good conductor of electricity and heat. Unless you're the type who enjoys a good hotfoot, *don't wear shoes with nails.* Rubber soled boots or shoes are all right, but somehow welders often seem to step into oil or grease, which dissolves the rubber and makes it more slippery than leather. The best welder's boots are

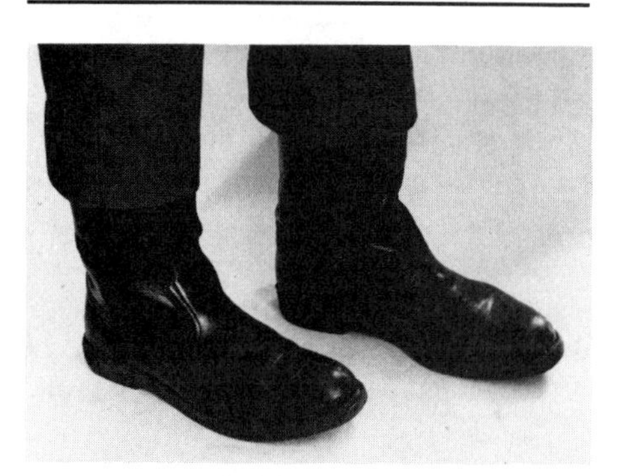

FIGURE 5-15.
Welder's boots. These boots must have soles without nails, a closed top, and a safety cap of steel in the toe.

waterproof, have a steel cap inside the toe, a neoprene sole, and *no nails*. Wet shoes or wet feet around electricity mean *trouble!*

A welder also needs tongs or vise grips, a wire brush, and a chipping hammer. One look and you'll know

FIGURE 5-14.
The leather apron also helps keep red-hot sparks away from your body.

FIGURE 5-16.
A wire brush and a chipping hammer. These tools come in many different styles, and they are necessary for cleaning welds.

why the latter is called a tomahawk. They make a fancy model tomahawk with the head composed of several layers of metal that slide independently. It looks strange, but it actually knocks slag loose twice as fast as the other kind. If your shop has lots of compressed air and money there's an air powered gadget with a fistful of loose rods sticking out of one end that can remove more slag in a minute than you can in an hour's worth of pecking away with a tomahawk. But you'll want ear plugs while using it. Whether you're swinging a tomahawk by hand or using an air tool, you'd better be wearing safety glasses. The clear lens type is just fine for deflecting jagged chunks of flying slag.

## Safety Precautions

### ARC-WELDING

Here's a review of the most important items to remember about arc welding:

1. Before starting the welder, check to be sure it's grounded and that all leads are safe to use.
2. Wear safety glasses at all times.
3. Never look at the arc with the naked eye.
4. Wear proper clothing.
5. Never carry matches where a spark can find them.
6. Do not wear cuffed trousers.
7. Wear proper shoes.
8. Make sure your welding helmet is in perfect condition.
9. Provide screens or other protection for other workers in the area.
10. Be sure the welding area is well ventilated.
11. Remove all flammable materials from the welding area.
12. Never weld on closed containers until they are made safe for welding by thorough venting or by filling with inert gas or water to limit the volume available for an explosion.

Precautions that are important to other types of arc welding will be discussed as they appear. Electric arc welding is not dangerous if safety rules are observed.

# 6 ARC WELDING

## THE ARC WELDER

The exercises that follow are necessary if you are to attain certification as a welder, and they are presented in the order in which they should be taught. Since some students may have no experience at all in arc welding, we shall start at the absolute beginning.

Because each make of welder is different (Figures 6-1 and 6-2), your instructor will show you how to set the arc welding machine. Even after you're an expert welder, setting amperage is largely a matter of trial and error.

It might be well to point out here that some authors attempt to distinguish between people and machines by calling the machine a weld*er* and the person who uses it a weld*or*. These confusing attempts at changing a perfectly good word do no harm, since both words are pronounced the same. But since most readers of this book are more interested in welding than spelling, this sort of hair-splitting does little good.

## THE ELECTRIC ARC ROD

There are many different welding rods used in the electric arc welding process. Basically, these rods look pretty much alike (Figure 6-3), but the metal alloy of the core or the composition of the coating can vary a great deal. The coating burns with the rod and forms a gas shield that protects the weld zone from oxygen and nitrogen in the air (Figure 6-4). These substances, should they enter the weld, cause brittleness and porosity. The protective gas shield is what gives arc welding its technical name: Shielded Metal Arc Welding (SMAW).

The bare end of the rod is inserted into the welding electrode holder, which is commonly called the stinger.

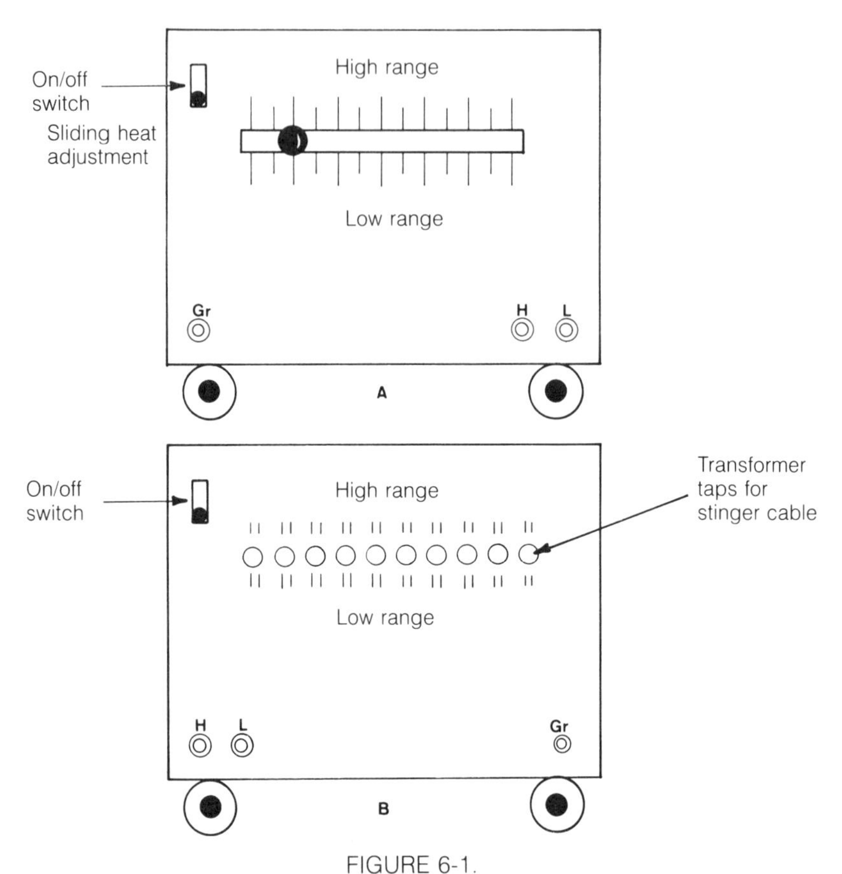


FIGURE 6-1.
Two types of AC welding machines. With machine A the welding current is raised by sliding the adjusting knob. With machine B the current is controlled by inserting the lead or electrode cable into the desired tap.

Spring tension in the jaws of the electrode holder holds the welding electrode firmly so the electric current can pass freely through the rod. Some of the more popular rods used in industry today are listed here along with the purpose for which they may be used.

E 6010—A very popular rod, used a great deal in the welding of pipe and other pressure vessels.

E 6011—The companion rod to E 6010, this may be used on pressure vessels with AC current if DC current is not available.

E 6013—A rod used mainly for arc welding thin sheet metals. Penetration is very slight.

E 7014—A heavily coated rod used mainly for fabrication and production work. Coating consists of about 50% iron powder.

E 7024—A very heavily coated rod used in production work. Coating consists of about 60% iron powder and can be used in only two welding positions: flat and horizontal.

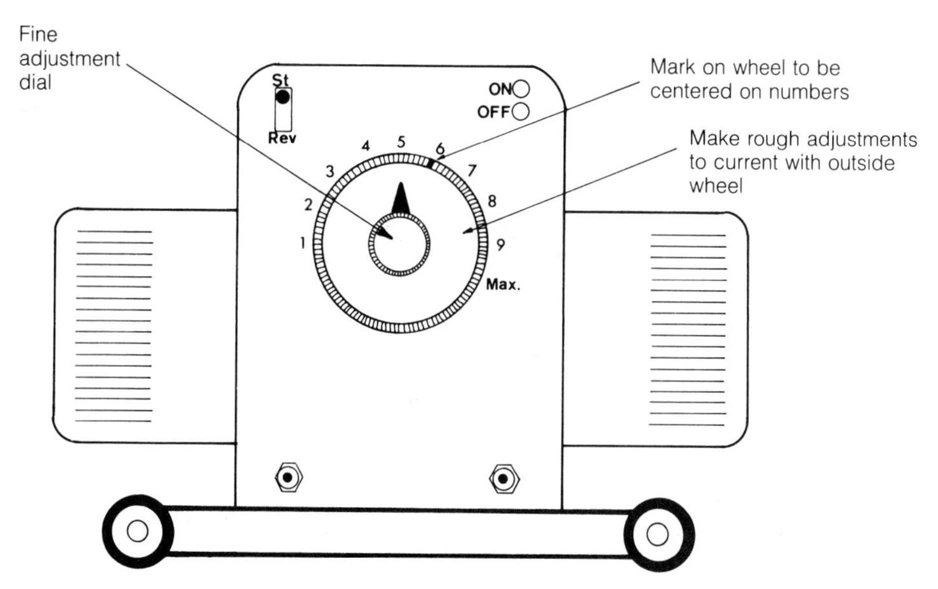


FIGURE 6-2.
A motor-driven DC arc welder. The large hand wheel is used to make rough current adjustments, and the small hand wheel is used to make finer adjustments.

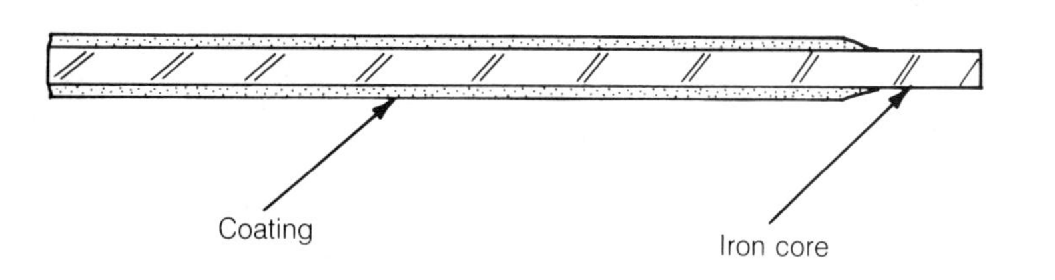


FIGURE 6-3.
The basic construction of all shielded-metal electrodes.

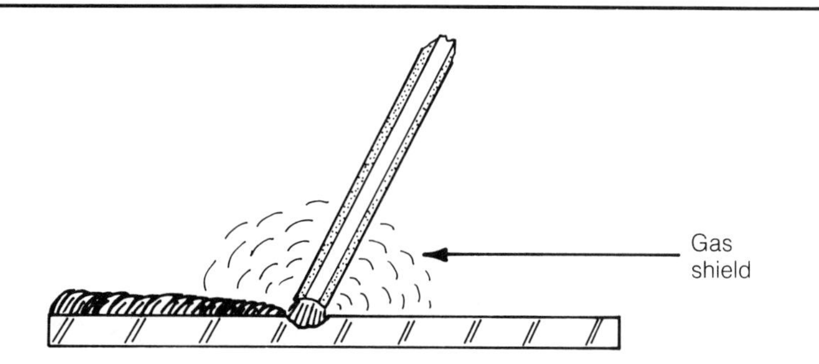


FIGURE 6-4.
When the coating burns from the shielded-metal electrode (rod) it forms a gaseous shield around the arc. This gaseous shield excludes the nitrogen and oxygen in the air from the weld zone.

FIGURE 6-5.
A welder striking the arc.

E 7018—Probably the most popular rod in use today. It is a low hydrogen rod, used in the construction industry, shipyards, and wherever a high tensile strength weld is needed.

Remember, the color of the coating on a welding rod has nothing to do with the composition of the iron core, and there is no standard color for the coatings. The alloy number will be printed on the box and sometimes on every rod.

## STRIKING AN ARC

Your first exercise will be with E 7014 rod, and calls for a plate of steel 1/4″ thick, 4″ or 5″ wide and about 10″ long. E 7014 rod strikes and holds an arc easily, which will help you acquire confidence. Begin on the left side of the plate and weld across the short way (opposite direction for left handers). First, pose with the stinger in position, then nod to lower the hood. Try not to move your hand. Then, when your hood is down, touch the rod to the plate. The arc will strike and begin to melt the electrode. Practice striking and stopping the arc a few times before trying to run beads.

FIGURE 6-6.
Stringer beads made by a beginning welder. Notice that the beads are not consistent in size or uniformity. This is normal for a beginning welder and can only be improved by practice.

## STRINGER BEADS

After the student has learned how to strike and hold an arc he or she should practice making stringer beads across the plate. This first exercise should be done in the flat, or downhand, position. Figure 6-7 shows the rod angle in relation to the plate.

***Flat Position.*** The angles shown in Figure 6-7 are only approximate, since the amperage, type of rod, and position of welding all have an effect on these angles. At first, each bead should be run singly, without touching the preceding bead. Once you have learned to form good, smooth beads of even size with good penetration, discontinue using E 7014 rod and change to 5/32" diameter E 6010 and repeat the same exercise.

### When the Rod Sticks

Two new things must be learned this time. First, striking the arc will be more difficult because this type of rod tends to stick, and students sometimes panic when this happens. The whole length of rod starts to heat up like the element of an electric stove.

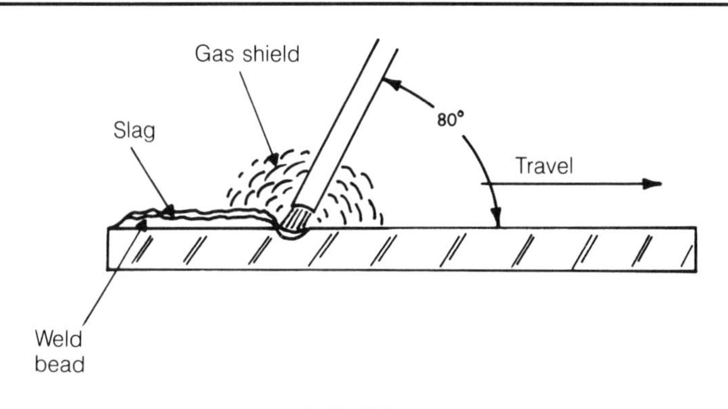


FIGURE 6-7.
Approximate rod angles for running flat stringer beads.

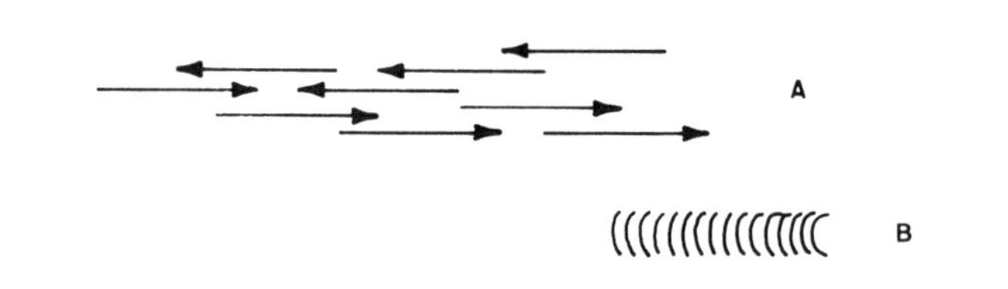


FIGURE 6-8.
The back and forth oscillation pattern of the rod is shown in the first illustration (A). If done correctly, the bead should appear as shown in the second illustration (B).

The coating starts to smoke and burn away, and within seconds the rod is red hot. Trying to tear it loose is hard on gloves and harder on hands. Try wiggling the stinger, and if that won't break the stuck rod loose, loosen the stinger and pull it away, leaving the rod stuck to the plate. This rod stays hot a long time so break it loose with pliers or tomahawk—*not with your hands*. And remember, when pulling the stinger away from a stuck rod, there will be arcing. Look away or close your eyes to prevent flash. Throw the rod away. With the coating burned off, it's no good any more.

Sticking is caused by jabbing the rod into the work. Instead, move it sideways as if striking a match. This movement is why we say "striking" an arc.

## Oscillation

The second thing you must learn about E 6010 rod is to keep it moving. Oscillation means to keep the tip of the rod weaving back and forth across the weld pool. This keeps the molten metal agitated and helps the slag and impurities float to the top where they can be removed by chipping and brushing. As the weld progresses, quickly move the rod forward about 3/8″. As the weld pool solidifies, move the rod back. Continue this back and forth movement as shown in Figure 6-8.

Your instructor will grade these welds strictly. Being basic exercises in arc welding, they are very important, and your instructor probably will have to demonstrate frequently. Before you progress to the next po-

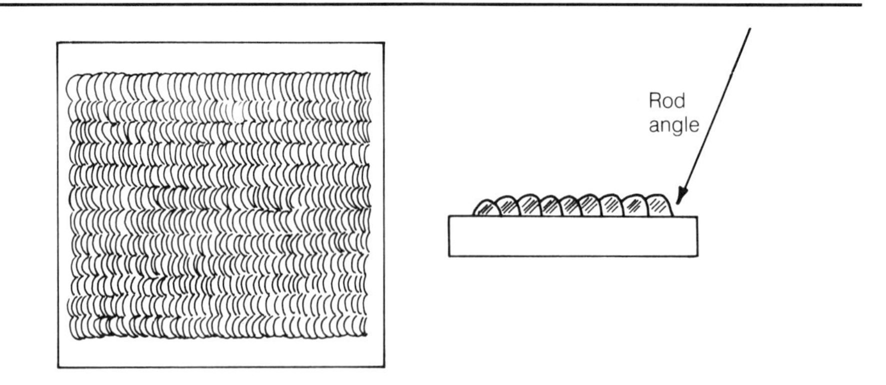


FIGURE 6-9.
This is how a well-formed pad of beads should look after you've finished.

sition, look at Figure 6-9, which shows what a flat, downhand, weld pad should look like. Note that, unlike your first practice beads, these must overlap.

There should be no flaws such as uneven beads or lack of penetration, and each bead should be tied perfectly into the preceding one (Figure 6-10). This leaves a pad of weld that is smooth and even, with no slag inclusions. In grading these pads, the instructor will decide when your skill is acceptable for industry and construction standards.

FIGURE 6-10.
This is how your flat pad should look after welding. Notice how each bead is securely tied into the preceding one.

These overlapping weld pads should be done not only in the flat position, but they should also be done in the vertical, overhead, and horizontal positions. Some welders would say that when you can build pads like these in any of the welding positions, you can do whatever you want, or need to do, with a welding rod. This is only partially true, since other factors enter the picture.

***Horizontal Position.*** After perfecting the flat, or downhand, position, the student should advance to the horizontal position. This position is not difficult to learn, but it's important to learn the correct way to weld in each position. In the horizontal position, the weld is applied onto a vertical surface, with the weld axis running horizontally as in Figures 6-11 and 6-12.

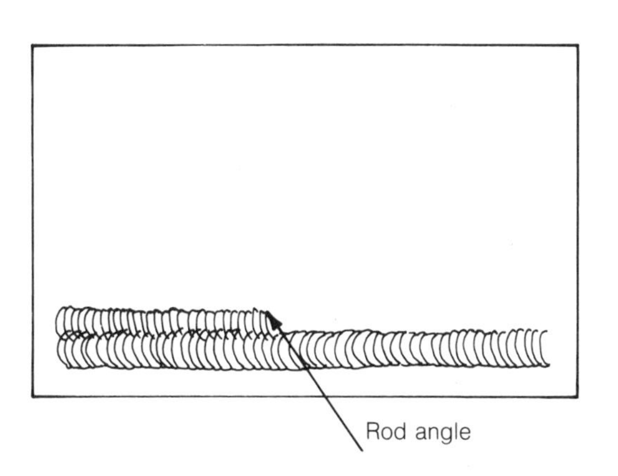


FIGURE 6-11.
Rod angle for the horizontal weld pad. These angles are not constantly the same. They may need to change very slightly from time to time.

Again, the rod angles are only approximate and may change slightly as the weld progresses. Notice in Figure 6-13 that two different angles are indicated when applying this weld.

Start at the bottom of the vertical plate and work upward. Apply the rod just below the top of the weld pool to achieve a smooth pad. Your instructor will probably have to demonstrate frequently and pay strict attention to your progress. This arc welding position is more difficult to learn than the flat position, and it will take more time, but if you are to be certified, you must master all four positions perfectly.

FIGURE 6-12.
Approximate rod angle for a horizontal weld.

By now you should be properly oscillating the welding rod almost automatically. You must strive for uniformity without undercutting and attempt to keep the weld level. Keeping the weld level should not be taken lightly. Most students, after running six or eight beads, will finish with the bead running downhill. This is wrong, and it should not be accepted by your instructor. If you develop this fault you must practice and learn to correct it. It is one of the harder things you have to learn, almost as difficult as it was to learn to write straight without lines on paper, but it *can* and *must* be learned.

After you have learned to weld the horizontal pads you may think you know enough about heat control to put arc weld metal wherever you want, but, as the next exercise will show, there are still things to be learned.

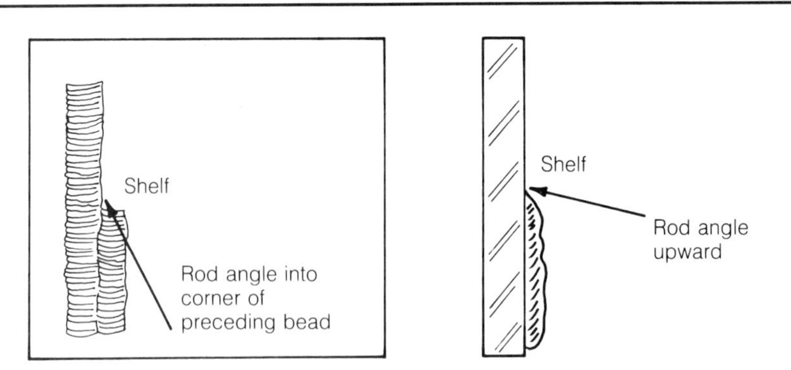


FIGURE 6-13.
The vertical weld bead. Notice the upward angle of the welding electrode and the angle into the toe of the preceding bead.

***Vertical Position.*** This is probably the most difficult position, and you will do much welding in actual shop work in this position, so it's important for you to take the time to learn the vertical position well. You *must* be able to show good arc control while welding in this position.

The plate is placed in a jig so that the surface is vertical, as with the horizontal position, except the weld should run from bottom to top. This requires the formation of a small shelf on which the weld can be placed. Figure 6-13 illustrates this technique.

Oscillation of the rod, rod angles, and your coordination are all important factors. Angle the rod upward ten to fifteen degrees. The force of the arc will help to keep the molten metal in place. This position is not easy. It takes time and practice to get the feel.

Timing is *most* important. As the rod is moved upward, about 3/8″ away from the weld pool, watch the portion of the weld that forms the shelf. As soon as this portion solidifies, move back into the existing crater to form the next shelf. Timing must be perfect in order to make the pad smooth and even.

This timing of the oscillation is one of the most difficult things a welder must learn. If the time out of the pool is too long, weld that must be welded over will be deposited ahead of the bead, and this will cause a rough, dirty bead. If the timing of your oscillation is too short, the last weld pool will not solidify and the weld will tend to run off the surface of the plate. This will cause an extremely rough surface on the weld face. The timing of the oscillation is very important. Time in and time out of the pool should be roughly equal.

This is a point in your training where you may become discouraged. Some students learn easily, but for others it is extremely difficult to achieve the required coordination. Welding is not a trade that can be learned overnight. In many ways it's

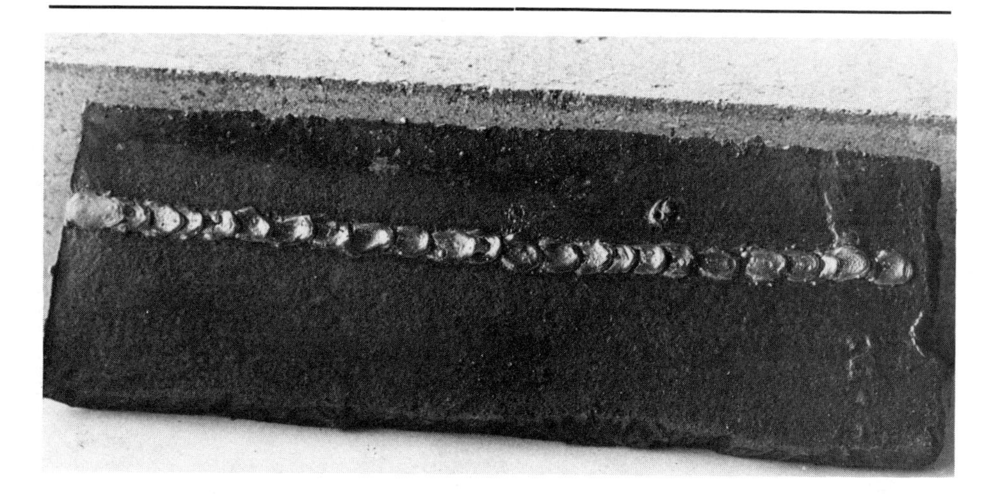

FIGURE 6-14.
The results of uneven timing in the oscillation of the electrode on a vertical weld bead.

more of an art than a trade, and to become a good welder can take as many hours or years of constant practice as to become proficient at tennis, skiing, or any other endeavor that requires fine coordination of mind and body.

Practice is necessary on any welding position, but the vertical position is the hardest to master. Advancement to the overhead position should come only after you have perfected the vertical position. For your own safety and well-being, your instructor should not be lenient when grading your vertical welds.

***Overhead Position.*** At this point, you should remember the previous warning about the danger of falling sparks from overhead welding. *All* the dangers of overhead acetylene welding are present when arc welding. But arc welding sprays more and hotter sparks. Proper clothing is necessary. Leather jackets or sleeves should be worn for protection.

Place the plate in the jig with the face to be welded facing downward. (See Figure 6-16.)

The welding machine will probably have to be readjusted. The amount of readjusting depends on the student's ability to hold a close arc. The amperage used for any given size electrode always depends partly on your ability to do a smooth job, and some students will need more or less heat than others. There *is* a rule of thumb which says to use one ampere of heat for each thousandth of an inch of rod diameter. In other words, a ⅛″ rod, converted to its decimal equivalent of .125″, would use approximately 125 amps. Again, this is a "rule of thumb," and it will vary according to the individual welder, the composition of the rod coating, and the welding position.

The actual overhead weld, however, is not as difficult as it may seem. Actually, it's only a flat, or downhand, weld done upside down. The rod angles are the same except that

FIGURE 6-15.
These little streaks of light are sparks and molten metal from the overhead weld; they can be described in one word—HOT! Molten metal falling on the welder may reach a temperature of 3000°F.

they are directed upward instead of downward. It'll be slightly harder for you to hold a steady arc in this position because of the weight of the electrode holder, the cable, and the fact that you are working in an unnatural position, but it's really no big deal, and a little practice will go a long way.

The weight of the cable and electrode holder may be relieved by hanging the cable from a hook or holding the cable with your free hand to take the weight off of your welding hand. Hanging the cable over your shoulder can be dangerous. If there's a leak in the cable insulation you might get a shock or, more probably, a burn.

Continue practicing on the overhead plate until you can consistently lay beads across the plate tightly against each preceding bead, while leaving the pad smooth and even. For some this will be easy. For others it will take long, hard hours of practice; however, this is a certification test position, and it must be learned well.

## DEEP GROOVE OR VEE JOINTS

The following exercises teach you the art of welding in a deep groove or Vee. The construction of this joint is

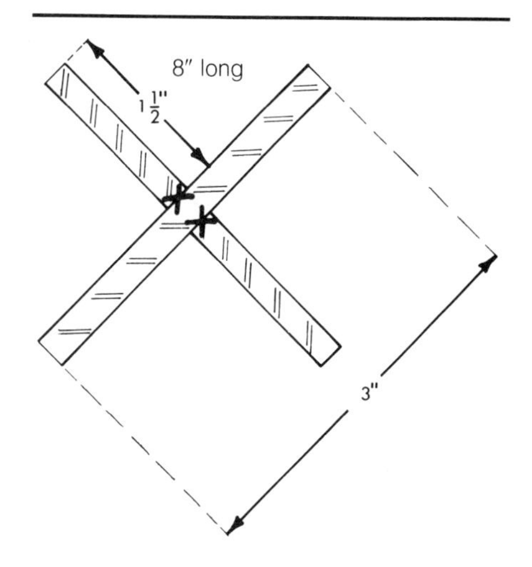


FIGURE 6-17.
The deep Vee groove weld is assembled from three pieces of steel sheet metal. Tack weld on each end at the points marked with an X. This will help you avoid welding over a cumbersome tack placed somewhere in the middle of the assembled plates.

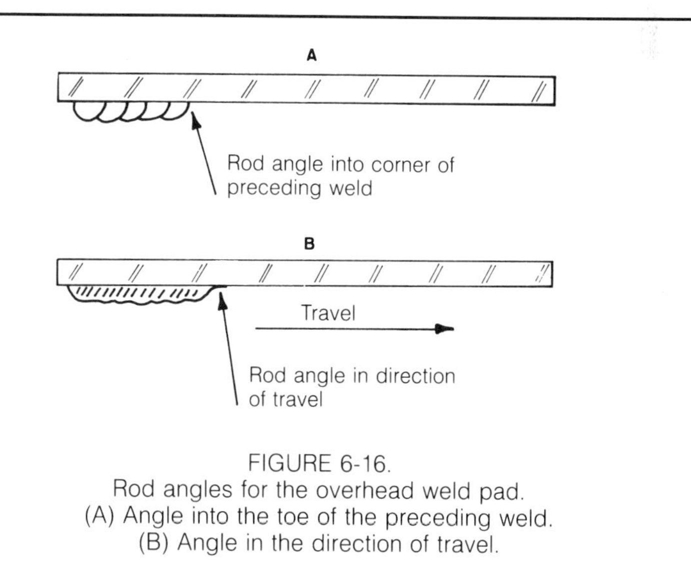


FIGURE 6-16.
Rod angles for the overhead weld pad.
(A) Angle into the toe of the preceding weld.
(B) Angle in the direction of travel.

FIGURE 6-18.
A properly assembled deep Vee groove joint.

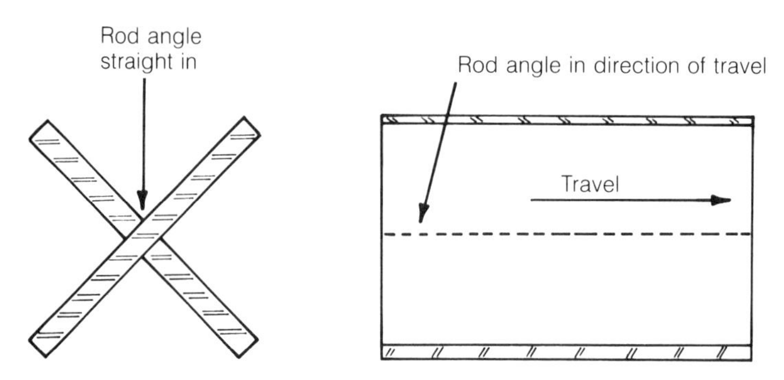

FIGURE 6-19.
This is how the deep Vee weld is placed in position for welding. Notice that the rod angle heads straight into the joint in the direction of travel.

shown in Figure 6-17. It is made of two pieces of 3⁄16″ steel 1½″ wide and 8″ long, and one piece of steel 3⁄16″ by 3″ by 8″.

***Flat Position.*** The two small pieces are tack welded to the large one to form a cross. Figure 6-19 shows the approximate rod angles for this exercise.

E 6010 rod is about right for welding these angles. Place the joint flat on the welding bench as shown in Figure 6-19. If the rod is changed from this straight downward angle the bead will try to climb up the plate on one side or the other depending on which way the rod is pointed. The root pass, or first pass, must be evenly divided on both plates.

FIGURE 6-20.
A deep Vee weld with the final cover pass applied.

Because this will be a multiple pass weld (several beads run on top of one another), each corner must receive a root pass before applying the second or third passes. When the hot seam cools it shrinks and pulls the parent metal with it. Any weld on the inside of an angle will close that angle. In construction this must be compensated for by hammering, or, if possible, by deliberately setting up the job with the angles open too wide. You will soon learn to "eyeball" how much a weld is going to shrink and can set up your work accordingly. This shrinkage applies to every dimension. A finished assembly is always smaller than it was before welding.

When applying the second pass, the center of the rod should remain directly over the edge of the first pass, and again, the rod must be directed straight into the joint being welded. The third pass goes on the opposite side, and the weld, with three stringer passes, should be smooth and even with no undercutting. A stringer pass is a weld drawn the entire length of the joint without a weave pattern. Use this same procedure in welding all four corners.

FIGURE 6-21.
The box, or basket weave, pattern. Keep this motion in the form of a box "C". Cross the face of the weld rapidly but keep the weld pool intact. Pause on each side of the seam to lay more metal along the edges to prevent undercutting.

After welding three or four joints the student should be able to lay the rod evenly, and the same joints can then be used to practice the box, or basket, weave. This is a cover pass,

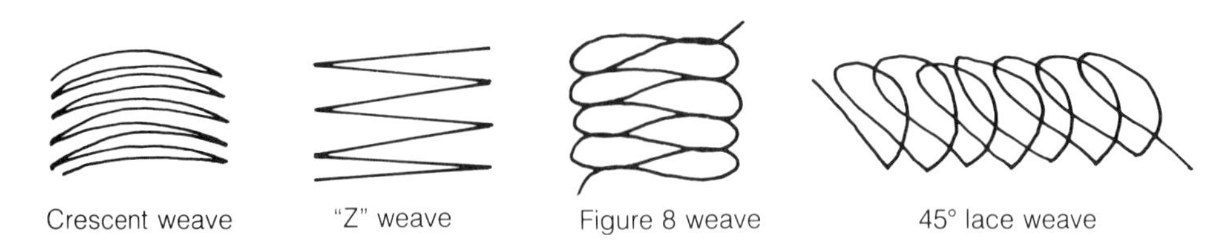


FIGURE 6-22.
Other cover pass patterns that may be used.

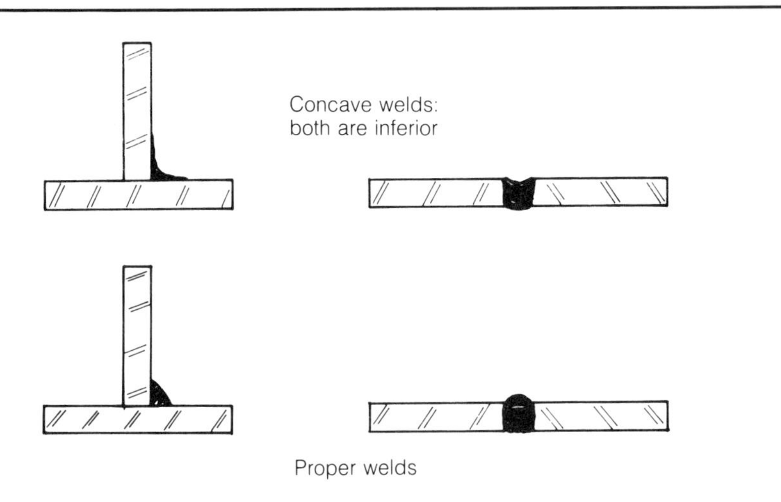

FIGURE 6-23.
The two welds at the top of the illustration may be classed as inferior welds. The finished shape of the welds does not conform to standards usually required, nor will the welds have the needed strength.

or cap, as it's sometimes called, that is used to smooth out any irregularities in the stringer passes.

## The Cover Pass

After finishing the three stringer passes the seam should be about ½″ wide, smooth, and even. This flat-faced weld gives the student a good base upon which to learn weave passes. Figure 6-21 shows the box, or basket weave, oscillation pattern. Figure 6-22 shows other patterns that can be used. The box weave, however, seems to be the easiest.

Take care not to go any farther from side to side than to the edges of the weld already deposited by the stringer beads. If the cover pass goes beyond the edge of the stringer passes the weld will appear concave, which is objectionable (Figure 6-23). Also, this may cause undercutting along the edge of the weld.

This cover pass, or cap, will be used in all four positions of this exercise. It must be mastered because it will be a part of your testing for certification.

If the face of your weld has holes

FIGURE 6-24.
The results of arc blow. Notice the excessive splatter on the plates surrounding the weld.

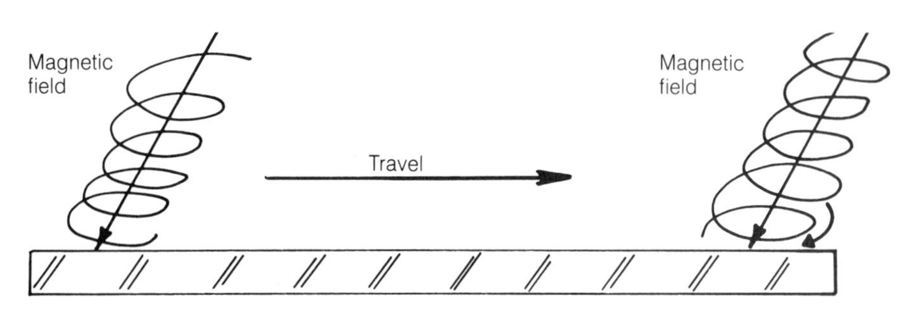


FIGURE 6-25.
Arc blow—the arc welder's worst enemy! The magnetic field around the welding rod is attracted to the base metal, causing the arc to assume an erratic path. The only sure cure is to use an AC welding current.

or slag inclusion, you are welding too fast—slow down! If you are moving too slowly, the weld will pile up in the center and become rough surfaced. A demonstration by your instructor is necessary here, and you should pay close attention. Note also in Figure 6-21 that a definite pause is necessary at each side of the weld to make sure the edges are filled and no undercutting or slag inclusions are incurred.

The first four joints that you welded with stringer passes should be used for learning this weave pass in the flat position. The vertical position will be more difficult to perfect because the weld will tend to run off the face. Here, you will probably encounter "arc blow" (Figure 6-24).

## Arc Blow

Arc blow is caused by the flow of electricity through the welding electrode (rod), and only happens when using DC. Except under very extreme conditions, there is no arc blow with AC. Figure 6-25 shows what causes arc blow.

The higher the voltage and amperage used, the stronger the magnetic field around the rod; thus, with large diameter electrodes, arc blow becomes a major problem. As the weld progresses toward the edge of the plate, the magnetic field tries to remain on the plate. This pushes the arc out of line, making a rough and unacceptable weld.

There are many suggested solutions to this problem, but the only real way to stop arc blow is by changing to AC. This is not always possible, however, so the student must learn other tricks. Changing the angle of the rod can help, but the best thing to do is to hold a close arc, so the current will not have a chance to wander. If the length of the arc does not exceed the diameter of the electrode, the chance of arc blow will be lowered.

***Vertical Position.*** The student must take great care with rod angles while welding deep Vee joints in the vertical position. Two angles are important. First, the rod should bisect the plates directly, without veering. Second, as the weld progresses, also hold an angle about 10° upward. This is difficult because as the weld progresses upward there is a natural tendency to change the angle. It should, however, remain constant throughout the weld.

A number of vertical weldments should be made with three stringer beads, as on the flat weld. At least six or eight of these joints will be needed. As mentioned before, this is a very difficult position to master, and again, it is a position that will be

FIGURE 6-26.
Proper rod angles for the vertical deep Vee weld.

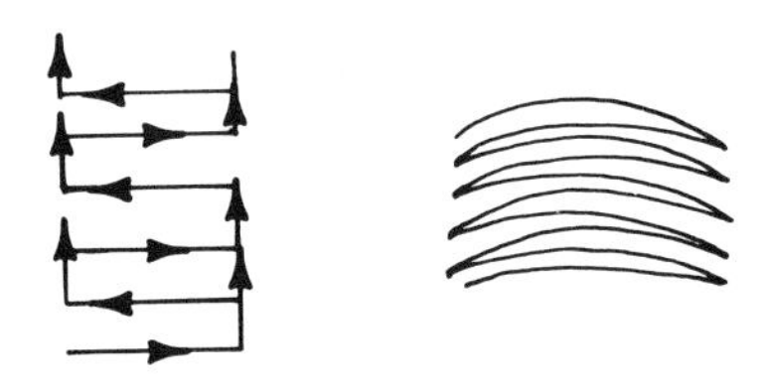

FIGURE 6-27.
Cover passes for the vertical position.

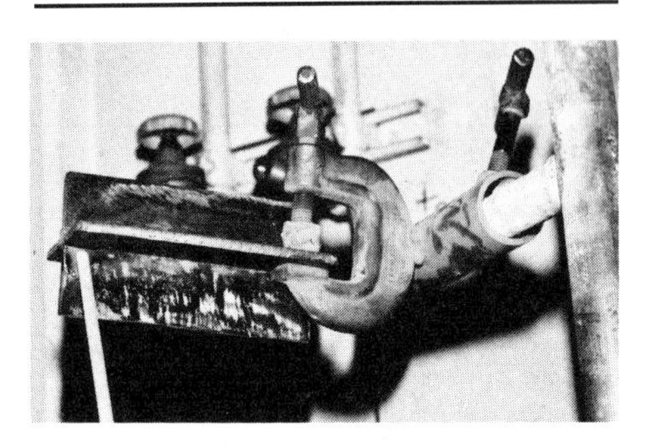

FIGURE 6-28.
Approximate rod angle for the overhead deep Vee weld.

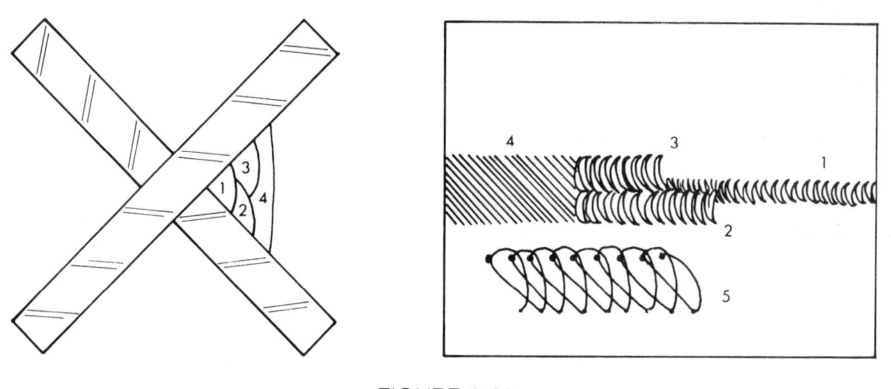


FIGURE 6-29.
Welding sequence for the horizontal deep Vee weld.
(1) root pass
(2) second stringer pass
(3) third stringer pass (watch for undercutting here)
(4) cover pass
(5) the proper rod motions for the cover pass.

given on certification tests. The stringer beads in this position may not give you much trouble but the cover pass must be applied very carefully.

Beginning at the bottom, after the three stringer beads have been made, form the shelf from which the cover pass or cap will be made. The diagram below will show the necessary movements of the electrode for a smooth, neat weld. Pauses at the edges are important to prevent undercutting and slag inclusions.

It will take more than one rod for the upward cover pass. Be sure to clean the weld crater thoroughly when changing rods. This will remove slag and prevent inclusions.

***Overhead Position.*** Welding deep Vee joints in the overhead position is the most difficult. Here, the weld is exactly opposite from downhand. The rod angles are opposite, but the progression is the same. This weld can be finished two ways: with all stringer beads, or with a box weave cover pass. Certification tests do not require the final box weave cover pass, and you *may* find this weave pass difficult to master quickly. Your instructor will decide whether you ought to learn the final weave pass or finish the entire weld with stringer beads. The motions and rod angles in Figures 6-28 and 6-29 should be studied carefully to prevent undercutting. The

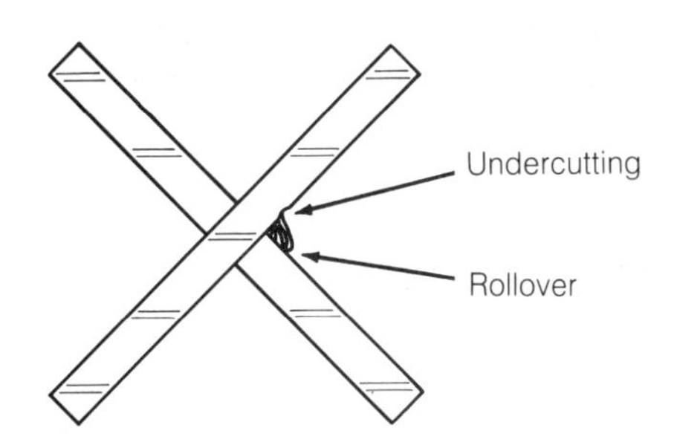


FIGURE 6-30.
Undercutting and rollover on a horizontal weld bead. An excess of heat and/or improper rod angle or speed of travel may cause this condition.

technique shown works for both overhead and horizontal positions.

***Horizontal Position.*** In the horizontal position you must watch the molten weld pool to make sure the weld at the top of the third stringer pass is being filled properly and not leaving an undercut. Liquid always flows down, and the vertical surface of this weld is no exception. It takes a lot of practice to perform this weld without undercutting at the top or rollover at the bottom (Figure 6-30). To prevent undercutting, rod angles will have to change slightly as the weld progresses.

The most drastic change of position will be in the way the cover pass is applied. See Figure 6-31. Sometimes it's called a 45° lace, other times it's just called a cover pass. Either way, it's difficult. Pause longer at the top than at the bottom when applying the weld. Only practice will tell you how long. Even though this position is not usually given as part of the certification test, it should be learned thoroughly.

***45° Deep Vee Joints.*** After you have mastered the horizontal Vee joint, your instructor may want you to practice 45° deep Vee joints. This weld can be omitted if your instructor wishes, but you should be able to weld with either hand, and that is the reason for the following exercise.

Place one weldment on a plate and tack it at a 45° angle to the right. Tack another weldment on the other end of the plate at the same 45° angle, only this time place it to the left. Weld each of these joints in a fixed position, the one to the left with the left hand and the one to the right with the right hand. Learning to weld with the opposite hand is not difficult. See Figure 6-32.

## CERTIFICATION PRACTICE

The next exercise deals directly with certification testing. Certification tests are usually given on 3/8″ steel plates 5″ wide and 8″ long.

***Flat Position.*** For practice the welder may use plates slightly smaller in overall size, but they should not be thinner than 3/8″. To save steel you might want to practice on a plate cut down to 3″. Leave the plate length the same so you can become familiar with the length of the seam which must be run. See Figure 6-33 for details.

A backup bar about 2″ wide will be sufficient for practice, although a 3″ wide, 3/8″ thick backup bar will usu-

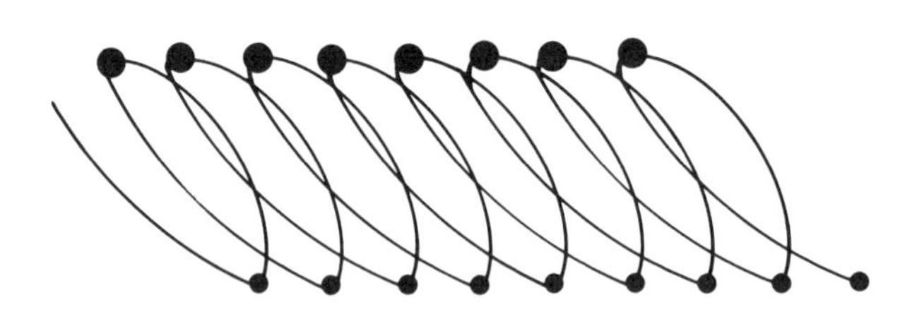

FIGURE 6-31.
The proper cover weave pattern for the horizontal deep Vee weld. Large dots at the top represent a longer pause than the smaller dots at the bottom.

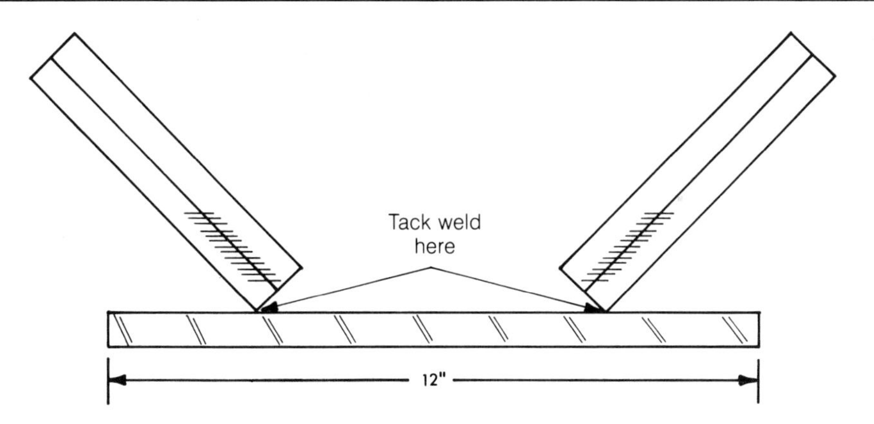


FIGURE 6-32.
Setup for right- and left-hand welds (45° Tee joints). Weld the left side with your left hand and the right side with your right hand. It's easier than you think.

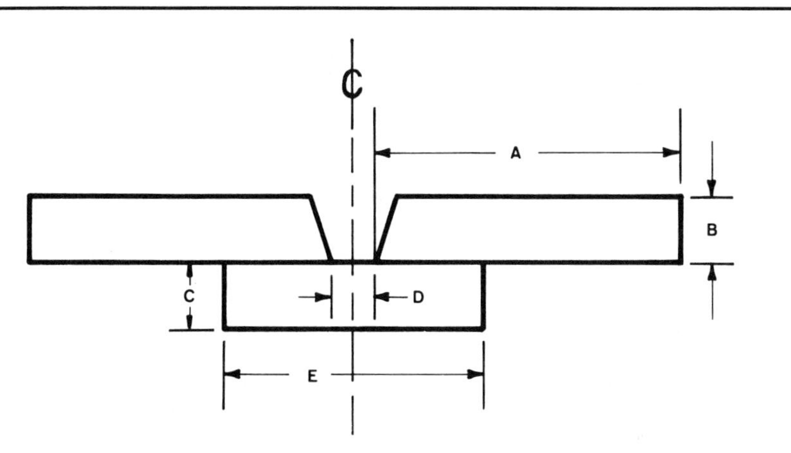


FIGURE 6-33.
The setup for welding certification tests. These dimensions will change according to the agency administering the test, and your instructor will have to fill in the missing numbers.

ally be required for certification tests. To aid in avoiding arc blow it is suggested that the backup bar extend about an inch beyond the ends of the plates being welded. At least two plates should be welded flat, or downhand, to make sure the student knows how to apply each pass before starting on the other positions.

The first pass should do nothing more than weld the two plates securely to the backup plate, and form a good smooth base on which to make the rest of the weld. The second pass may be applied as a weave pass to make the filling process faster. As the weld material approaches the top of the groove, you should leave a definite line along the top of the beveled edge in order to apply the final pass. This final pass must be applied so that the weld does not rise more than 1/8" above the level of the plates being welded, or extend more than 1/16" be-

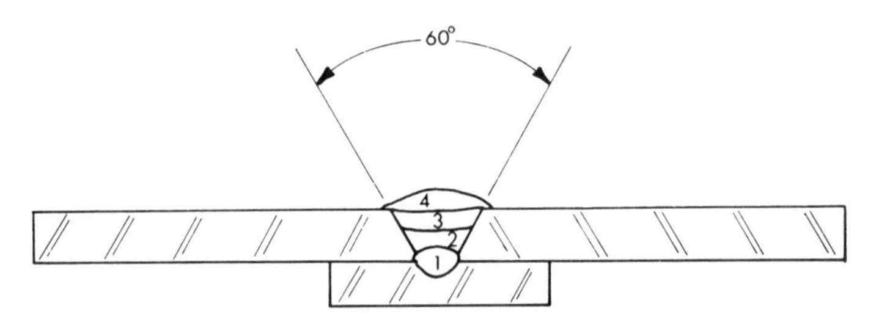


FIGURE 6-34.
Placement of beads on test plates. The size of each bead will determine the number of passes required to finish this weld. BE SURE TO CLEAN EACH WELD THOROUGHLY!

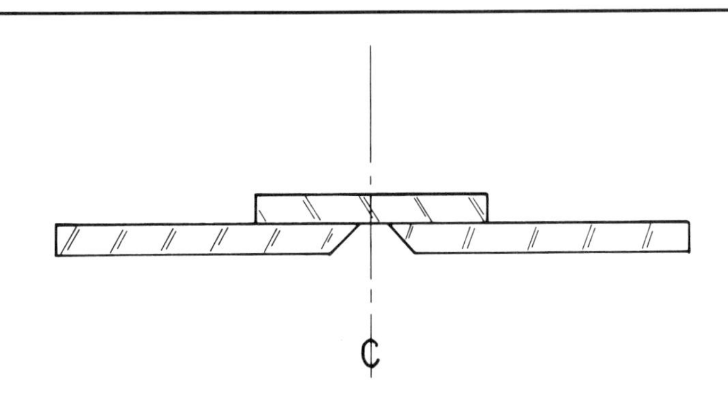


FIGURE 6-35.
Position of plates for the overhead certification test.

yond the bevel in either side. Testing laboratories are rather strict about this. They do not want a weld that is out of shape or one with excess weld where it is not needed.

***Vertical Position.*** The setup for the vertical weld is the same as that used for a flat weld, except that the position of the plates is different. The surface of the plates and the axis of the weld are both vertical. This position must be practiced to perfection, since it is one of the certification testing positions.

At this time there are two possibilities for certification testing: Your instructor may take you back to the flat position and have you make the same weld with E-7018 low hydrogen rod. This practice will probably be done on ¾″ thick plates. The tests themselves will be taken on 1″ plates.

Under the AWS Structural Welding Code a welder may take the certification tests on plate 1″ thick, using E-7018 rod, but, you may also take the test using E-6010 rod. Passing the test with E-7018 rod will qualify you to weld with any type of welding rod. If you pass the tests on grooved plates in both the vertical and the overhead positions you will qualify to make both fillet and groove welds regardless of plate thickness. This may be a great incentive to the student welder, but your instructor must help you decide whether, with a reasonable amount of practice, you could pass these tests.

The author tests his students with E-6010 rod first. E-6010 rod is easier to use than E-7018, and builds up the student's confidence. After students are reasonably proficient they may go on to the low hydrogen E-7018 rod.

While taking the vertical position tests, many students will use too much heat or try to put in too much metal on the first pass. This, of course, results in heavy slag deposits and other impurities in the weld. The purpose of the root pass is to tie the two plates solidly to the backup plate, and the only thing you should watch for is complete penetration. It will take about four or five passes to properly finish this weld, and the welder will have to use a slight weave. Clean each pass thoroughly with a wire brush before running the next. See Figure 6-34 for how to apply the last two passes.

It takes considerable practice to make these two final passes acceptable in appearance, and welding on grooved plates should be practiced until you can consistently run perfect beads no matter which way the plates are set up.

If you and your instructor both feel that you can pass these certification tests in 1″ thick plate with E-7018 rod, then you *should* take them with the low hydrogen rod. There is no point in taking two sets of tests when complete certification can be finished with one.

Note that more students fail the vertical tests than the overhead, because it's much easier to trap slag along the edges of a vertical seam than on an overhead seam.

***Overhead Position.*** Figure 6-35 shows the setup for certification in the overhead position.

On the first (root) pass the welder will have to use a definite weave to make sure the weld is penetrating into both edges of the beveled plates. ***Caution:*** Do not try to put on more metal than necessary to tie the two plates together. Be sure to watch for any possible lack of penetration. Too much weld on the first pass will cause slag traps along the edges of the beveled plates, and this is one of the major reasons for students' failures on certification tests.

Each pass should be thoroughly cleaned. Use a tomahawk or other

FIGURE 6-36.
The all-stringer-bead method of welding test plates in the overhead position. Notice that each bead is securely tied into the preceding bead.

sharply pointed tool and a wire brush to make sure all the slag and other foreign material has been removed—especially along the edges or toe of the weld. Each pass should be tied tightly to the last bead and, in order to avoid a lumpy weld, it should be the same size as your last pass. Lumpy welds take longer to finish and their appearance is undesirable. Appearance is one of the factors considered when you take your certification tests. Figure 6-36 shows the proper sequence for applying these beads.

***Horizontal Position.*** The horizontal position of welding on grooved or beveled plates is not a test position for structural welding as prescribed by the AWS; however, you should practice this exercise in order to sharpen your skills in the prevention of undercutting and the placement of weld beads.

In itself, the horizontal position is not too difficult to perfect. Rod angles are the most critical part of practicing this position. If the rod angles are incorrect, the weld will show a definite undercut. The amperage and the speed of travel across the plate are also factors that must be considered. These two items, however, are not as important as the rod angle, which must be watched continuously.

Set up the plates to be welded the same way you set up for the flat position. This same setup will be used in both the overhead and the vertical

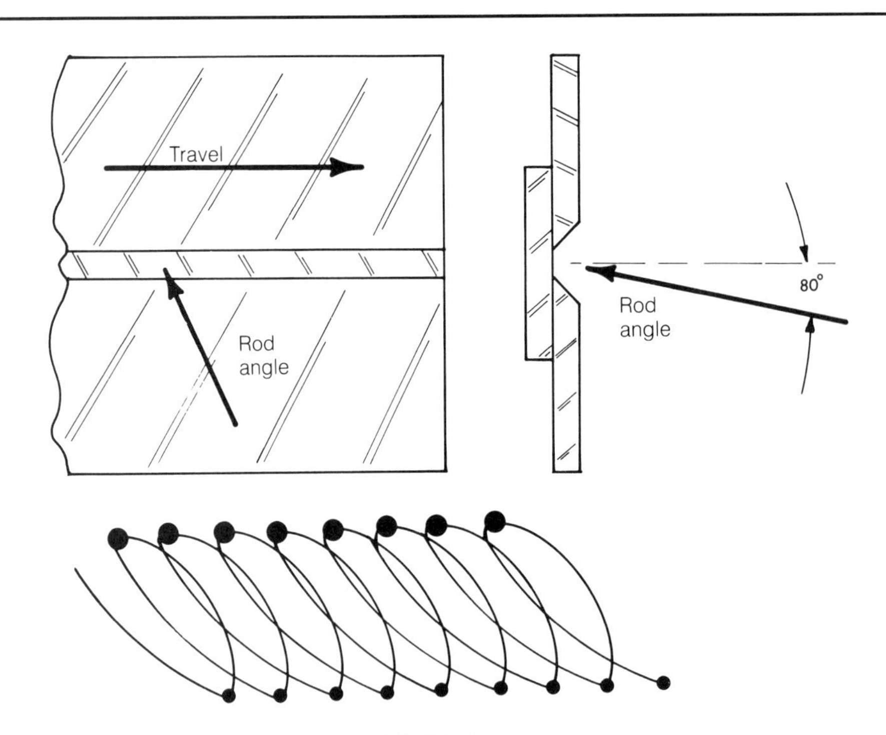


FIGURE 6-37.
Position of plates for welding grooved plates in the horizontal position. The rod angles and motion shown are for the final cover pass.

positions. Figure 6-37 should explain the setup and the position in which the plates are to be placed in the jig.

After the initial or root pass has been applied, you can finish this weld by using the all-stringer-bead method, or it may be done with weave passes. If the weave pass method is used, it must be applied at the angle shown in Figure 6-37. You should study the motions shown in Figure 6-37 and practice them until you have thoroughly achieved perfection in applying this weave pass.

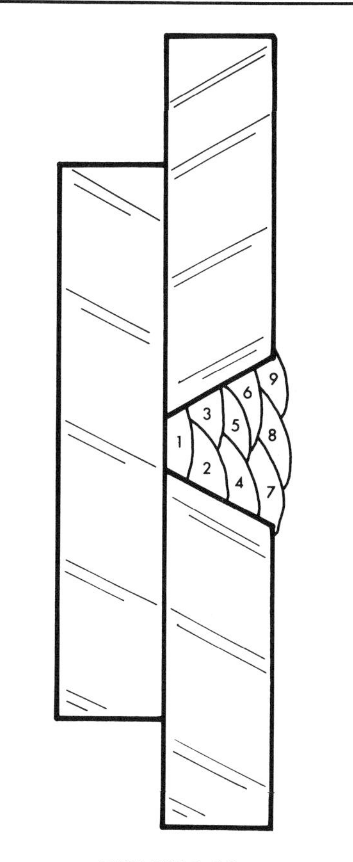


FIGURE 6-38.
Welding sequence using the all-stringer-bead method to finish a horizontal groove weld. Pay particular attention to the sequence of the weld pattern.

The easiest way to accomplish this weld is to use stringer passes all the way from the root to the final cover pass. Figure 6-38 shows how to apply the weld when this procedure is followed. Whichever method of application you choose, the rod angles are very important. From time to time they may change from the angles shown here, but, as a rule, they should remain fairly constant. Practice will give you the "feel" of when a slight change in this angle will be needed.

When you apply the root pass, be very careful to see that the edge of the bevel on the top plate is being welded solidly to the backup plate. If the rod angle is off, the weld bead

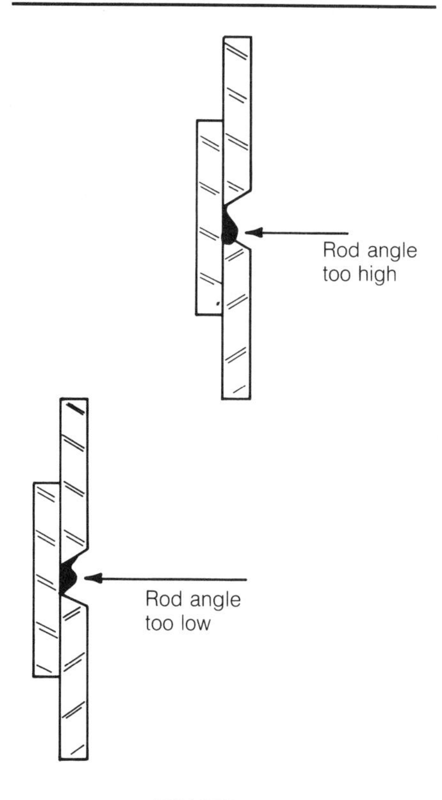


FIGURE 6-39.
The undesirable effects of wrong rod angle.

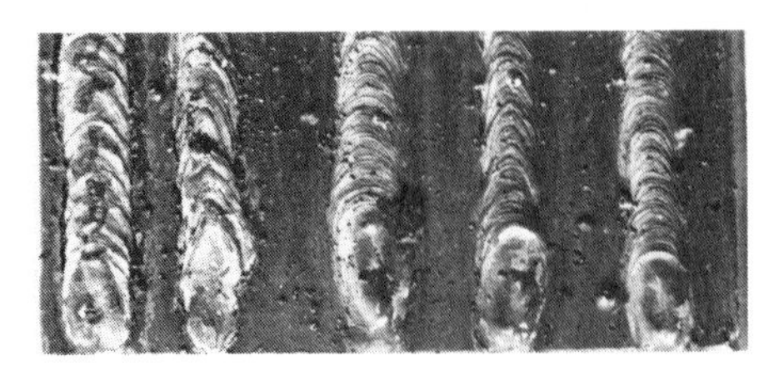

FIGURE 6-40.
Effects of wrong rod angle.

FIGURE 6-41
A guided-bend test machine. Samples are bent against both the root and face of the weld. Samples at least 1½" wide are cut and ground smooth—directly across the weld proper.

will have a tendency to roll downward. This will happen if the rod angle is too high or at a 90° angle to the root. If the rod angle is too low, the weld will bulge in the middle and

FIGURE 6-42.
Samples of welds that have been bent on a guided-bend test machine. The sample on the left has passed the test.

produce an undesirable effect. Figure 6-39 shows the effects of incorrect rod angles.

In all, this should not be a difficult position to master. To make a perfect weld, avoid an excess of heat, and watch your speed of travel closely.

Welding certification tests are expensive, and you should be very sure of yourself before trying to pass them. Your instructor should watch you carefully, and not let you go for certification until you both know you are ready. Practice increases confidence.

Evaluation is done in various ways, depending on the code under which tests are taken. Under some codes the welds are X-rayed, while other codes require a guided-bend test machine (Figures 6-41 and 6-42). Some tests even require a tensile strength or pull test (Figure 6-43). As a rule, however, these are special testing procedures and you should not be concerned about them.

FIGURE 6-43.
A tensile strength testing machine that pulls the weld sample apart.

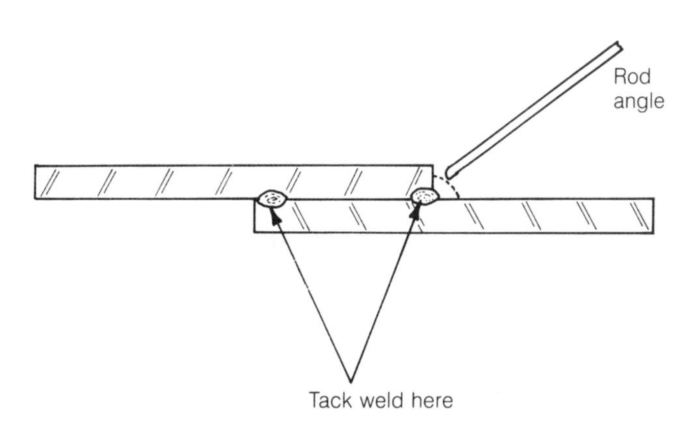


FIGURE 6-44.
The lap weld setup showing the approximate rod angle.

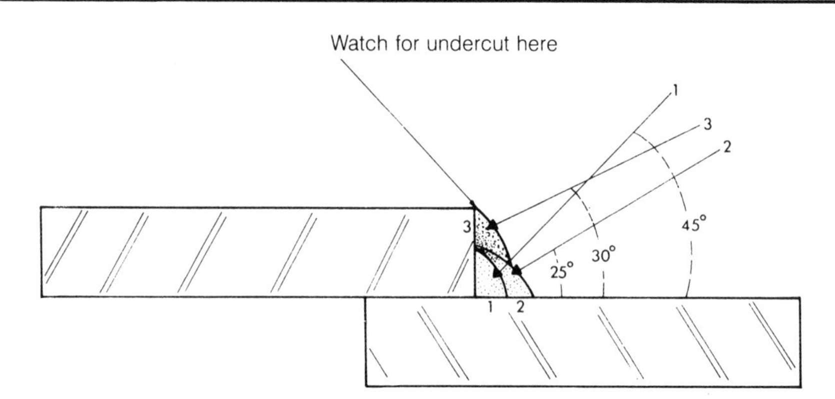


FIGURE 6-45.
Welding sequence and approximate rod angles for the lap weld.

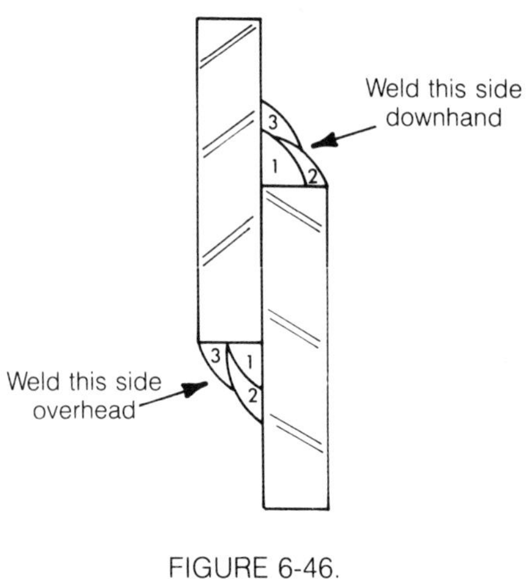


FIGURE 6-46.
The horizontal lap weld.

## OTHER WELDED JOINTS

After arc welding certification you should go on to other exercises that will help make you a good combination welder. The first can be the lap weld. This is not difficult. As in other welds, the lap weld should be done in all four positions and the results carefully studied to make sure they are done correctly. The lap weld is one plate lapped over another, as shown in Figure 6-44.

The major problem here is to keep from burning away the top plate. It's not hard in the flat position, but in other welding positions it's very difficult to keep from burning the top plate, especially in the horizontal position. You should always remember that it's easier and quicker to do it right the first time than to go back and patch up an undercut.

Two plates 1/4″ thick and about 6″ long are required to practice the lap weld. Place these plates as shown in Figure 6-44 and run the first bead at the bottom of the Vee formed by the plates. The first bead should go in at about 45° to the surface of the plates.

Don't make the first bead too large. This will raise the surface of the weld too high and make the second and third beads difficult. Apply the second bead so the center of the rod travels along the bottom edge of the first bead. This will leave a small shelf on which to lay the third and last bead of the weld. Take care here not to burn away the thin edge of the top plate. The speed of rod travel must be exact in order to fill this weld to the very top of the plate. See Figure 6-45 for the sequence of applying

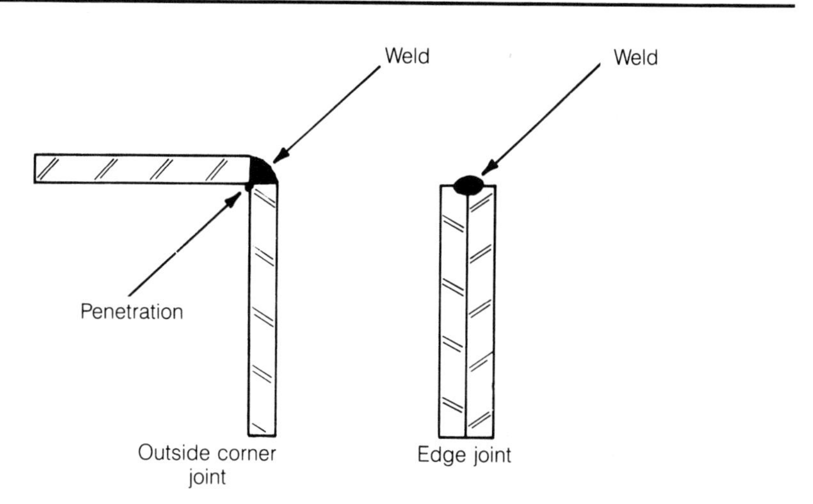

FIGURE 6-47.
The outside corner joint and the edge joint. A thin line of penetration should show on the inside apex of the outside corner joint.

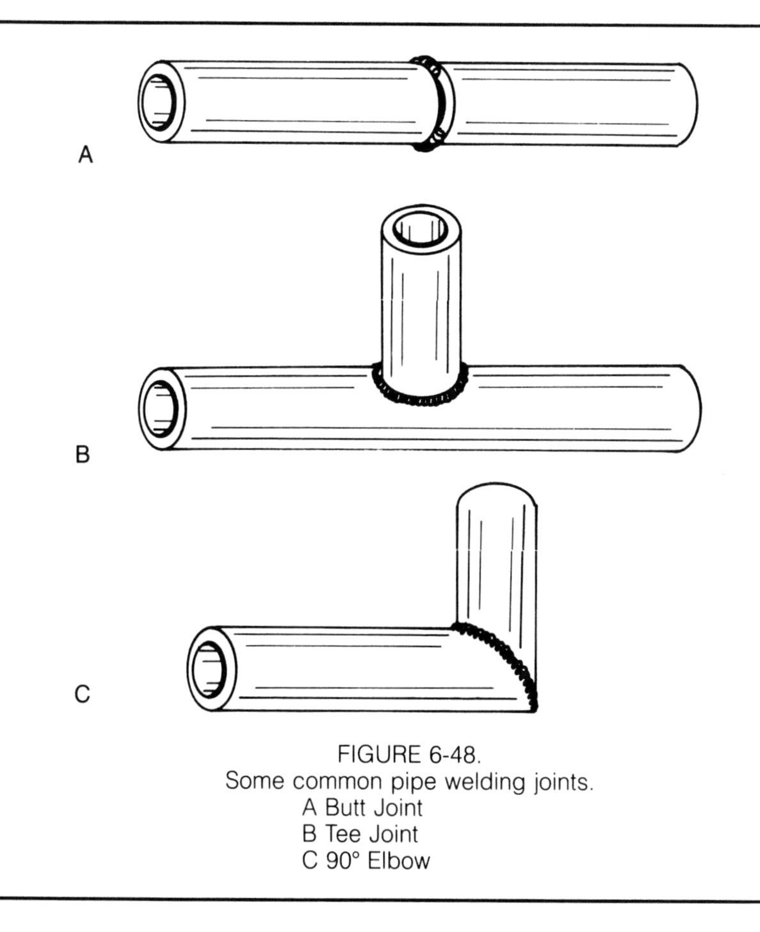


FIGURE 6-48.
Some common pipe welding joints.
A Butt Joint
B Tee Joint
C 90° Elbow

these beads and their appearance after completion. This weld gets easier with practice.

You may practice any other position as the next exercise. The author suggests trying the horizontal position, because it is a little more difficult than the flat position. Do it as shown in Figure 6-46.

One side of the plate should be welded from the underside of the joint and the other from the top. The underside will be more difficult, because too much weld will put an undesirable roll on the edge. In this position it's best to make a few more passes than to slop in an excess of metal that looks bad. The thin edge at the top of the plate will burn away very easily, so the amperage will have to be carefully adjusted.

Other joints that can be practiced after certification are the outside corner joint and the edge joint (Figure 6-47). These are not difficult, and they should be practiced in all positions. All the joints described in this book should be done with both thick and thin metals. Thin metals—less than 1/4" thick—are classified as sheet metal. This practice will give you the extra experience needed to apply for a job. Many welded products are made of thin sheet metal. These can be hard to weld, but a good welder should know how to do anything.

If you have time, try some pipe welding (Figure 6-48). Pipe welding is not required for certification under AWS structural codes, but it could help you get a job.

The student should never forget the importance of proper clothing. Welding is a dangerous trade and safety precautions must be observed at all times. **Welding machines don't cause accidents. Careless operators cause accidents.**

# 7 MIG WELDING

Semi-automatic (Metal Inert Gas or MIG) welding has been around for a long time, mostly in factories where mass production can justify the high cost of the machinery (Figure 7-1). The basic MIG process is a logical development of arc welding. By now most students of welding will know how time-consuming and annoying it is to be continually replacing rods in the electrode holder. A MIG machine does away with this by furnishing welding rod in a continuous roll, fed through a handle that has many and marvelous names, but is usually called a squirt gun.

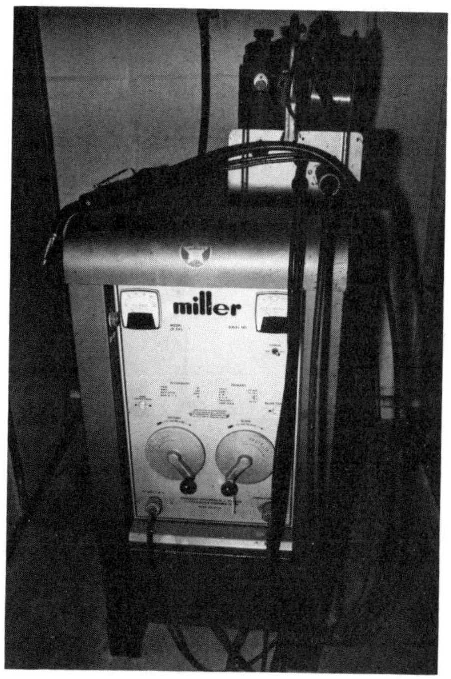

FIGURE 7-1.
A popular make of MIG welding machine.

The rod comes in rolls of wire fed through a complicated system of rollers whose speed can be controlled by the operator. Wire over 3/32″ in diameter is usually hollow, with the flux inside instead of outside, as with coated arc welding rod. This is called inner-shield rod. Smaller diameters of wire, from .030″ up to .068″ are not fluxed. With these small diameter wires the hot rod and molten pool of weld are protected from oxidation by a flow of inert gas that comes from the squirt gun to enclose the area being welded.

When welding steel, this inert gas can be carbon dioxide. Aluminum and other metals, however, are usually welded with a mixture of argon and helium. These gases are inert, which means they will not combine chemically with the metals being welded. In simpler words, they will not rust, corrode, burn, or support life.

MIG welding is actually easier than arc welding, since in arc welding one must keep moving the stinger closer to the work to make up for the length of burnt-off rod. With a squirt gun, rod is automatically fed to the weld at the same rate it burns off. Theoretically, that is.

Ninety percent of MIG welding, exercises, and setups are the same as for arc welding. The new things that you will have to learn are mostly concerned with heat settings, rate of wire feed, and mixtures of gases.

Since there are nearly as many makes and models of semi-automatic machines as there are welders, this book will leave it to your instructor to explain the settings and controls of whatever machine is being used in your training situation. Though mysterious at first, they're not all that hard to learn. Figure 7-2 shows a typical MIG welding setup.

## GASES

Gases used in MIG machines range from mixtures of argon and oxygen to pure gases. Non-ferrous metals are often MIG welded with argon-helium combinations. Helium is light and tends to blow away, but it conducts electricity and produces a hotter arc, thus welding highly heat-conductive

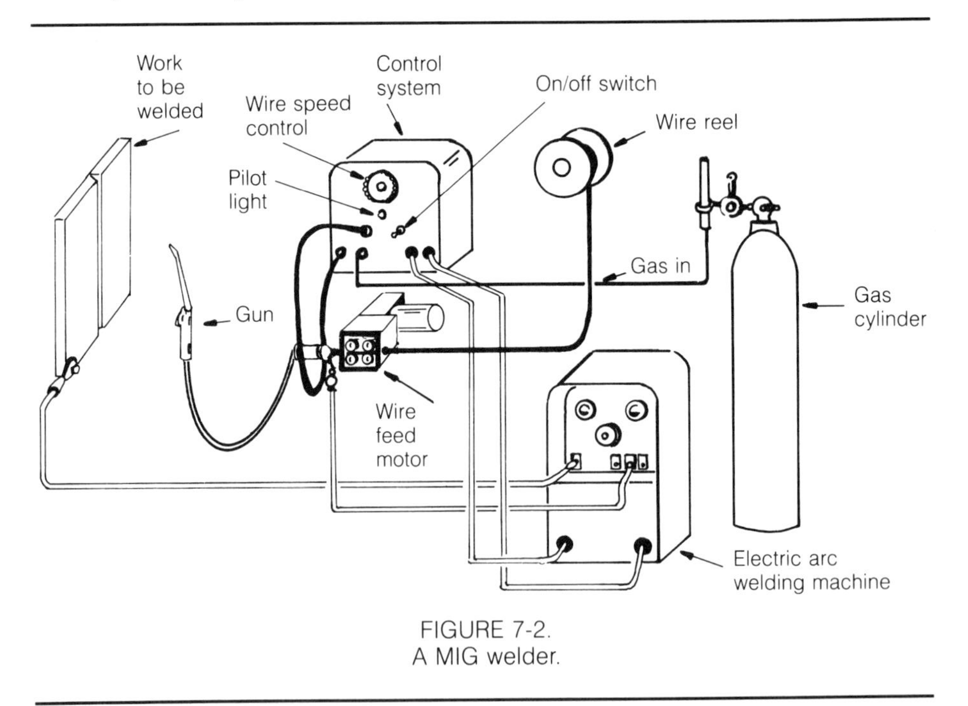


FIGURE 7-2.
A MIG welder.

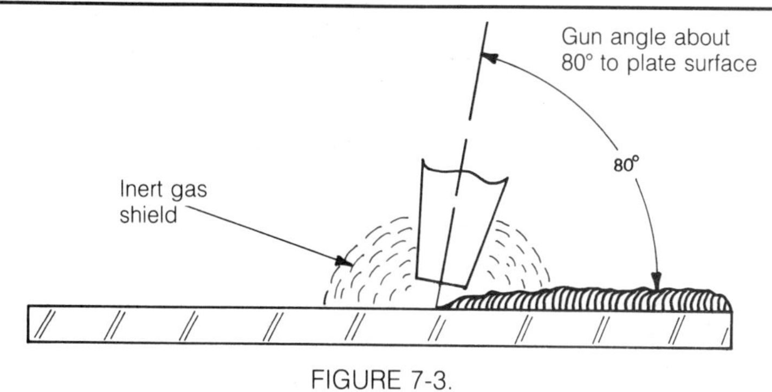


FIGURE 7-3.
The proper gun angles for flat welding with the MIG welding outfit.

metals (aluminum or copper) with lower currents. Where heat loss is not important, heavier argon gas is substituted for helium.

Carbon dioxide ($CO_2$) is much cheaper than the other gases and seems to work well on most of the mild steels in use today. For high carbon steels $CO_2$ must sometimes be mixed with the more expensive inert gases.

Carbon dioxide is usually stored in cylinders under high pressure, the 50 pound capacity cylinder being the most popular. Larger storage units may be used where the volume of work warrants the extra capacity. A 50 pound cylinder will supply about 435 cubic feet of gas, and a full cylinder will have a pressure of about 850 psi. At 70°F about 35 cubic feet per hour will be released from this cylinder.

It is necessary to use a slightly higher amperage with $CO_2$ than with some of the other gases, but with today's mild steels the penetration gained is well worth it.

$CO_2$ will leave slightly more spatter around the weld than other gases, but again, this disadvantage is nothing compared with the advantage of penetration and sound welds.

FIGURE 7-4.
A quench tank used to cool metals after cutting or welding.

Gas mixtures for other types of welds and exotic metals are largely learned as a result of experimentation and experience. For the time being it's best for you to concentrate on getting the feel of the machine.

## ARC LENGTH

Welding with a short arc length permits welding with a lower arc voltage and is always easier when using DC. Since an AC arc drops to zero voltage many times per second, the arc tends to be less stable, but an AC arc is less subject to arc blow so it all averages out. Experienced welders will often "choke" (shorten) an arc just to save a long walk back to adjust the machine downward.

Arc length is especially critical in MIG welding. The rule of thumb is to keep the arc as short as possible but to avoid letting the electrode touch the parent metal. This contamination causes the arc to be erratic and the weld dirty.

## MIG STRINGER BEADS

After your instructor has demonstrated how to set up whatever machine is to be used, the first exercise will be running straight stringer beads. The purpose of this exercise is to teach you the correct angle and speed of travel used with the MIG welding gun (Figure 7-3).

Use a plate of mild steel 1/4″ thick, 5″ wide, and about 10″ long. This will be large enough to prevent the weld beads from sagging or flattening due to heat saturation. A quench tank (Figure 7-4) is helpful to prevent the plate from overheating. Amperages, and therefore heating, are much higher in MIG than in straight arc welding.

After completing the first pass, the second pass should be applied at the toe of the first pass. This will tie the beads together and make a solid layer of weld across the surface of the plate. Be very careful about the speed of travel across the plate. Your movements should be as smooth and steady as possible in order to make each bead the same width and height. You should practice until you can apply these beads smoothly; then go on to the vertical position. Your instructor should grade these welds very strictly, because they are very important in learning to do a good job with the MIG welder.

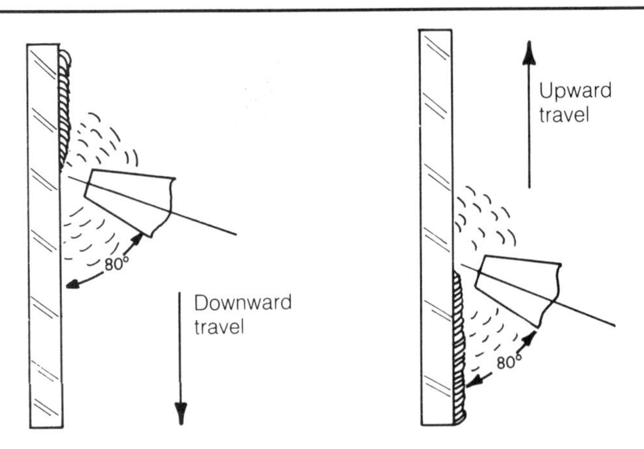


FIGURE 7-5.
Gun angles for both up and down vertical welding. Much less weld metal can be applied in a single pass when welding down.

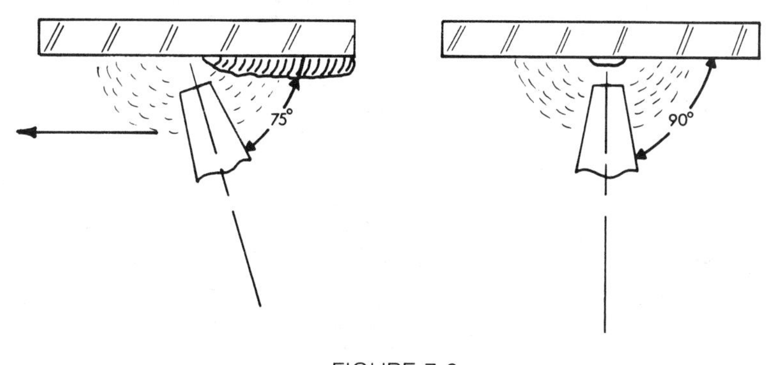


FIGURE 7-6.
Gun angles for overhead welding—MIG process.

***Vertical Position.*** There are two ways to weld stringer beads in the vertical position with MIG machinery, and both should be practiced to perfection. Use the same size plate that you used for the last exercise. Place it in a jig that will hold it vertical, with the face on which the weld will be applied straight up and down.

The most popular way of running vertical welds with MIG is to begin at the top and weld down (downbead). This gives a thinner bead on the plate, but it is practical because of the lack of slag inclusions. Amperage and wire speed may be reduced to give a larger bead, but when this is done penetration may be sacrificed. Study Figure 7-5 carefully for the correct angle of the gun.

Downbead welding should be practiced until the finished plates are smooth and even. Once again, the beads should be tied into each other to make a solid layer of weld across the plate. This is important in passing certification tests.

The other method of applying the vertical weld is to run the squirt gun from the bottom of the plate upward (Figure 7-5). Welding upward on the plate is a little more difficult because the molten weld metal has a tendency to run downward off the face of the plate. Therefore, it will probably be necessary to reduce the amperage to prevent this runoff. The tendency of weld metal to run off the plate will worry most students, and some will have trouble overcoming it, but "practice makes perfect," so practice until you can weld these plates smoothly and evenly. The vertical position is difficult, and takes a lot of time to perfect.

***Overhead Position.*** Overhead welding will seem difficult for a while, but it becomes easy with a little practice. See Figure 7-6 for the proper angles.

The greatest danger from welding in this position is the chance of getting a hot shower. Take care when aiming the gun; try not to stand where the sparks and molten metal are going to fall. Practice until all the welds come out smooth and even. This position is required for certification.

***Horizontal Position.*** Horizontal welding with the MIG welder is a little easier than overhead welding.

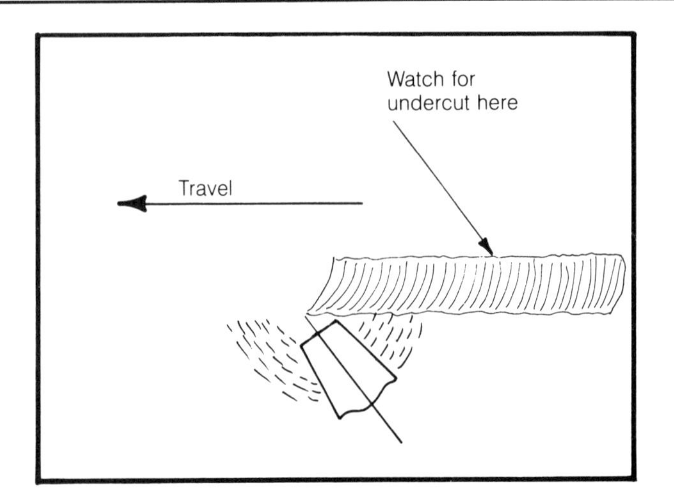


FIGURE 7-7.
Gun angles for the horizontal weld. Watch for undercutting at the top of the bead.

In this position, too, the plates should be vertical. Place the first bead on the bottom of the plate and work upward. Tie each bead into the last and keep the finished plate smooth and even. Torch or gun angles will vary as shown in Figure 7-7.

It is not necessary to sacrifice amperage for the finished plate to come out smooth and even. Your greatest problem will probably be keeping the beads straight and level. Most students will begin too wide or too close to the last bead, which will cause the pad to angle across the plate. A great deal of welding is done in this position so you should strive for perfection. The most important thing to practice on this weld is speed across the plate. A uniform speed keeps each bead the same size, and helps keep the finished pad smooth and even.

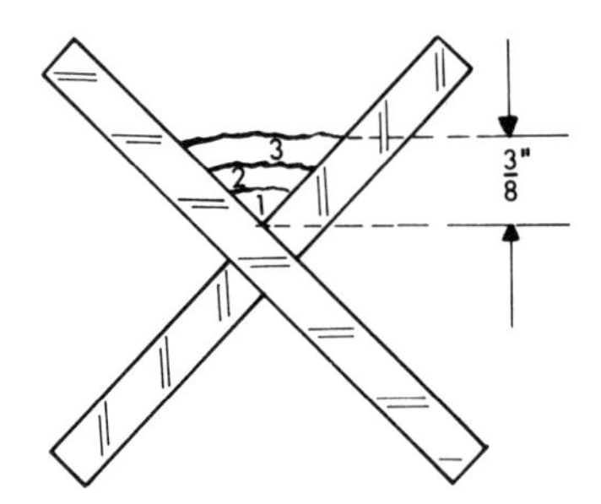


FIGURE 7-8.
Sequence used for welding the deep Vee joint. The three passes should bring the entire weld at least 3/8" from the root to the face.

## VEE JOINTS

***Flat Position.*** After learning to run stringer beads in the four positions, you are ready for Vee joints. Again, these will be done in the four positions, and the first position to practice is flat. This exercise is designed to help you become familiar with working in a deep groove and to position the gun. On your first pass you should use a stringer bead for the original root bead. Then use weave passes to build up the desired height. These tests are usually taken on 3/8" plate, so practice building up the deep

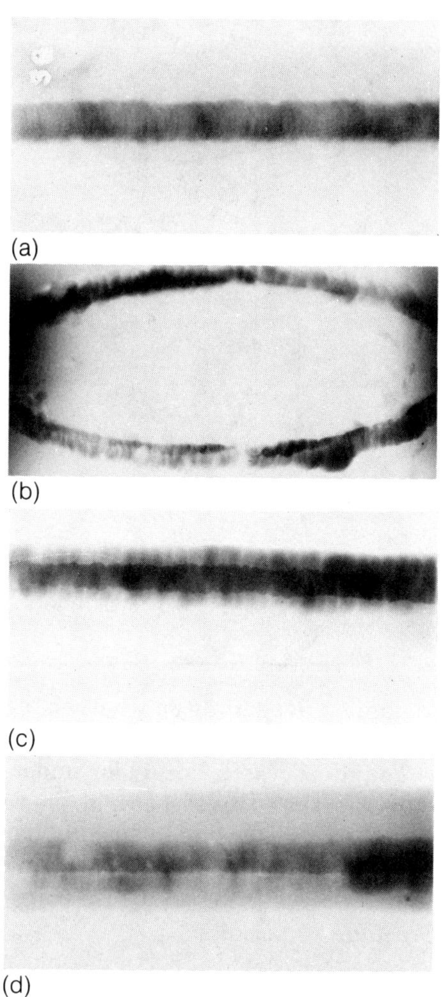

(a)

(b)

(c)

(d)

FIGURE 7-9.
(a) An X-ray of a perfect vertical weld. (b) An X-ray of a pipe weld that passed the certification test. (c) An X-ray of a weld that failed because it lacked fusion at the root of the beveled edge of one test plate. (d) An X-ray of a test plate that failed because of an extreme lack of fusion.

groove of the Vee joint to this height. See Figure 7-8.

Once you have started the weave pass you will probably find that the center of your weld is too high. This comes from not moving the gun from side to side fast enough across the weld face, and not pausing at each side before starting across. This timing will become *very* important in other weld positions of this type. Your instructor should grade these welds strictly. A thorough understanding of these timing sequences is essential in accomplishing the weld positions that will come later. **Do not try to put on too much weld at one time.** Too much weld with one pass will result in porosity and lack of penetration into the base metal. Figure 7-9 shows X-rays of two welds that passed the certification test, and two welds that failed because of lack of penetration.

***Vertical Position.*** MIG welding Vee joints in the vertical position can be simple even though it is difficult in the other welding processes. The easiest way to weld in the vertical position is downward, beginning at the top of the plate. If your gun travels across the plate too fast it will cause a lack of penetration. Too slow, and it will do the same thing by applying too much metal. Too much metal will make the weld run ahead of the arc and cover the base before it can penetrate.

After the first bead, you should use a weave pass to complete the weld. It will be necessary to use a fast whipping motion across the face of the weld in order to prevent building the weld center too high.

Once amperage and voltage are properly set you shouldn't have too much trouble with this position. It's mostly a matter of coordination. Figure 7-10 shows the approximate gun angles—try to stick to them. You will automatically tend to change the gun angle as you approach the bottom of the plate, and this must be avoided.

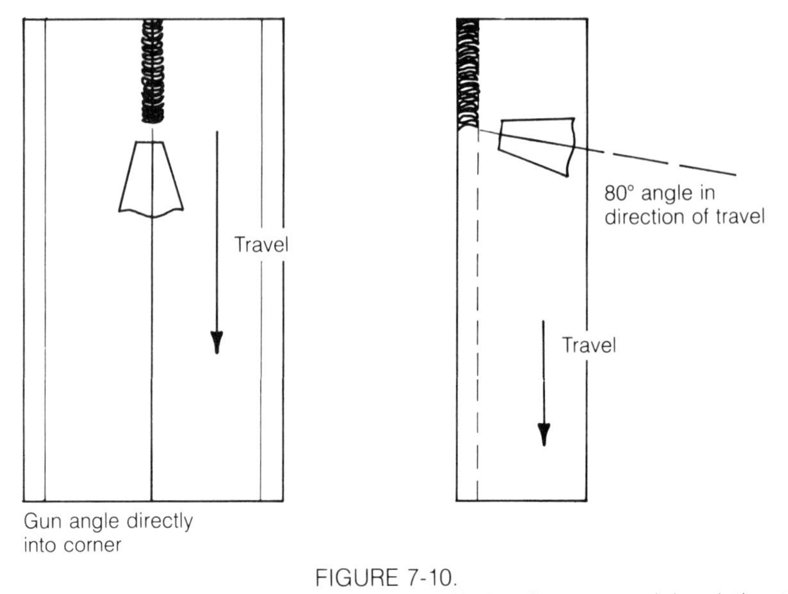


FIGURE 7-10.
Welding downbead vertical on a deep Vee joint. Notice the gun angle's relation to the plate.

Welding upward in the vertical position is entirely different from the downbead method. The amperage must be reduced to give you better control of the bead. Vertical upward is slower than downbead, but it gives better control of bead size and penetration.

Since this weld begins at the bottom and progresses upward, the face tends to bulge outward, especially on the first pass. The first pass is not too important, but on the succeeding passes you must try to keep the weld face flat. As the weave pass is applied to fill the groove, you must use a faster torch motion across the plate, while pausing on each side to prevent bulging. Appearance is a prime factor in passing the certification test.

Many students have trouble seeing the weld pool in this position. The weldment must be placed carefully in the jig so that you can see the entire weld puddle clearly, and it may be necessary to raise or lower the weldment depending on whether the seam is being run up or down.

You must watch closely for undercut along the edges of the weld. There is little danger of slag inclusions, but improper motion or holding the gun at the wrong angle will cause porosity and gas pockets.

***Overhead Position.*** Overhead MIG welding is more difficult than vertical welding, especially in groove or other deep angle welds. The two views of an overhead Vee joint shown in Figure 7-11 will explain the correct gun angles and weldment positioning for overhead MIG welding of deep angle welds. Follow these angles closely and do not change them as your gun progresses across the weld.

There are two equally good ways to make this weld. One is the weave pass, which is very difficult and dangerous. The sequence shown in Figure 7-12 is for a horizontal weld where

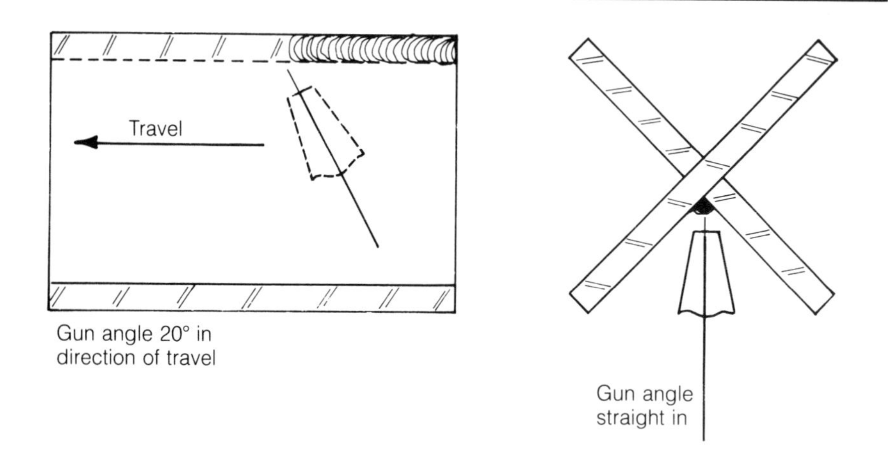


FIGURE 7-11.
Overhead welding in the deep Vee joint. Gun angles should be as close as possible to those shown.

all beads are run as stringers, but it applies to the overhead position as well. Each bead should be tied securely into the last so there will be no chance of porosity or lack of penetration. This is the easiest way to make this weld.

"What goes up must come down" applies to sparks and molten metal as well as other things. Wear leather sleeves or a jacket since, apart from hurting, these burns can cause infection.

When MIG welding in this position, the weight of the gun and cable is one of the hardest things to get used to. This is one reason for not practicing on plates longer than the 8″ required for certification. Practice until the weight of gun and cable becomes natural and a part of the job.

***Horizontal Position.*** Figure 7-12 explains the setup for welding horizontal Vee joints and shows proper gun angles.

Since there is little slag to contend with in the MIG welding process, welding these joints is rather easy. The weld will fail because of improper heat selection or a wrong gun angle, but little else will affect it. In any type of welding the proper rod and torch or gun angles are the most important factors in making a good, sound weld. This is especially true in MIG, TIG, and the oxy-acetylene processes. In the MIG and TIG processes these angles are necessary to ensure proper coverage of the weld zone by the inert gas. In the acetylene process, the outer envelope of the flame provides the needed cover.

In Figure 7-12, notice that the welding gun is made in such a way that the proper gun angles are very easy to achieve. Too many student welders try to form their own angles and end up in an unnatural position.

The welding gun is also made so that the cable and its wire liner should hang free of twists and turns while the welding operation is being performed. Under no circumstances should the welding cable be hung over your shoulder or from any projecting surface. These conditions will put an extra curve in the wire liner and may cause the wire to slow down in the feed or stop. When this happens the weld will be inferior.

The actual welding of horizontal Vee joints with MIG equipment is done the same way as it is done with the shielded metal arc process. You may weld it with the all-stringer-bead method, or you may apply a weave pass after applying a root pass. The gun motions should be the same as those used for arc welding, and the beads should be applied at an angle across the plate. Figure 7-13 shows the motions of the MIG gun while welding this joint.

If the handle of the welding gun is hanging straight down, the angle of the gun into the weld is correct. This is where a beginning welder will have a tendency to change the gun angle, and by doing so will obtain an inferior weld. If you maintain a comfortable position and use the equipment the way it was designed to be used, you should be able to weld with the MIG gun in any position. **Never drape the welding cable over your shoulders to relieve the weight of the gun.** Doing this will cause a malfunction of the equipment, or, if the welding transformer

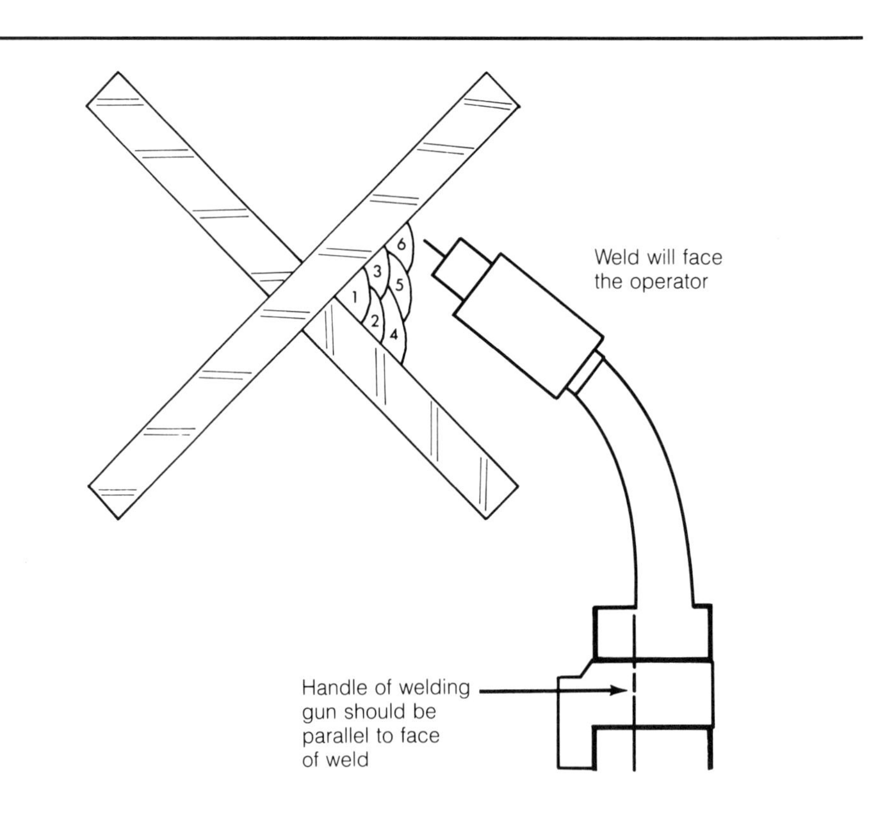

FIGURE 7-12.
The sequence for welding horizontal deep Vee joints. Notice the proper gun angle for horizontal welding.

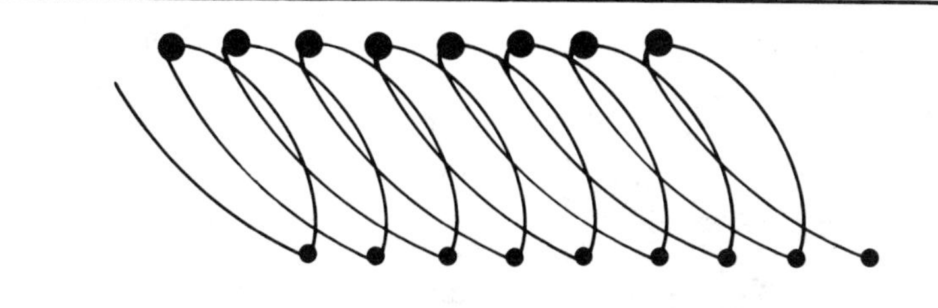

FIGURE 7-13.
Motion of the welding gun for weave passes on the horizontal deep Vee weld.

should short out, you might end up on permanent disability or on a slab in the morgue.

Under the structural codes this position of welding is not required on the MIG certification test. It is good practice, however, and once you've gone to work many welds will be done in this position.

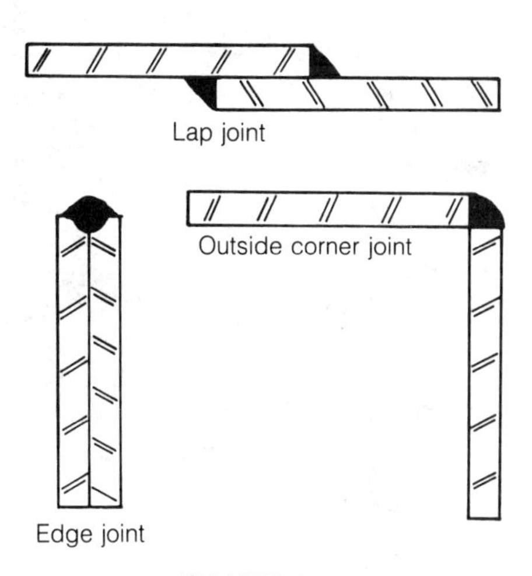


FIGURE 7-14.
Other joints that may be welded in all positions by the MIG welding process.

## OTHER WELDED JOINTS

Other joints that you must practice in the four basic positions are the lap, outside corner, and edge joints (Figure 7-14). Even though they're not included in the certification tests, your instructor will show you how to weld these joints in all positions. They will be used in a large part of your work after certification.

MIG is also used to weld thin metals, and you should practice welding and should learn the voltage and amperage settings for various gauges of sheet metal. A little experience in school will make it much easier once you are out on your first real job.

**Always remember to wear proper clothing.** The ultraviolet rays emitted by MIG welding are not shielded by smoke around the arc. The amperages are much higher than those used in arc welding, and your skin can be burned through a single layer of shirt. Ultraviolet radiation can do permanent damage to your skin or eyes, and the slightest gap at wrist or neck *can* cause skin cancer. Use common sense. When in doubt, ask.

## CERTIFICATION PRACTICE

The vertical and overhead positions are part of the certification tests (Figures 7-15 and 7-16). One plate will be welded downbead vertical and the other upward. Test plates are tacked together as in Figure 7-17.

The same plate assembly is used for both vertical and overhead certification by American Welding Society standards. Clean all plates thoroughly before tacking them together, and make sure there are no hollow spots under the beveled edges which can cause a lack of penetration in the root pass. Use the same procedures you practiced for the deep Vee joints with these test plates.

FIGURE 7-15.
A vertical plate, partially welded by the MIG process. Notice the wide root opening and the backup plate.

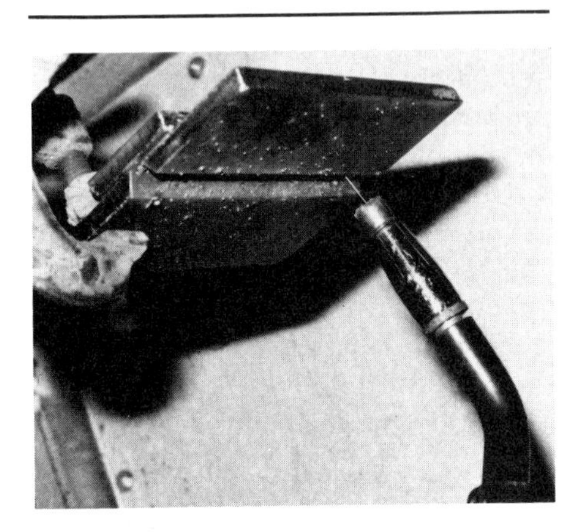

FIGURE 7-16.
The proper angles for overhead welding with a MIG gun. The spatter on the plates indicates that the welding machine is incorrectly set. Voltage and amperage must be correct to avoid spatter.

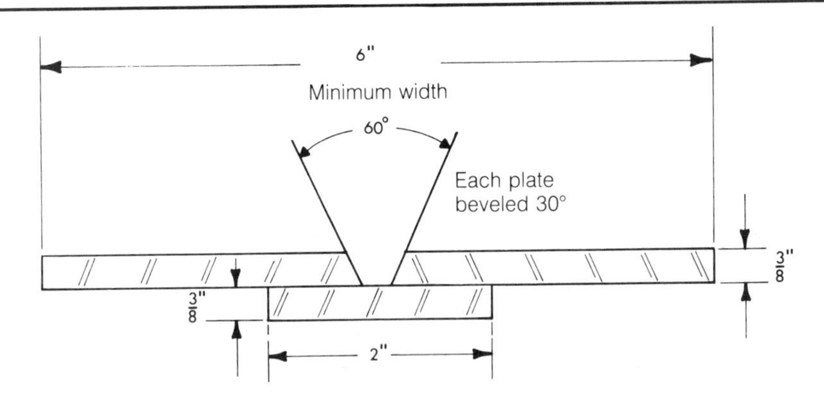


FIGURE 7-17.
Setup of plates for certification in the MIG welding process. These dimensions will vary according to the testing agency. Your instructor will inform you if they're different in your area.

# 8 TIG WELDING

Tungsten Inert Gas welding (TIG) is used extensively to weld aluminum, stainless steel, and other exotic *or* non-ferrous metals. Because it was first used with helium to join magnesium, old-timers still call TIG welding "heliarc." TIG is similar to oxy-acetylene welding in that the filler rod is separate from the torch and is fed with the opposite hand. The difference between oxy-acetylene and TIG is that the gases used for TIG are inert, as in MIG welding. The heat comes from an electric arc protected from the air by these inert gases. A TIG welding machine is pictured in Figure 8-1.

## TUNGSTEN ELECTRODES

The arc comes from a tungsten electrode, which is classed as non-consumable. This means the electrode will not burn up or deposit itself into the weld as with MIG or coated electrode arc welding. Tungsten is used because it is one of the few metals that will not melt at electric arc welding temperatures.

The first thing a student must learn is that the tungsten electrode should **never** contact the weld pool. If it does, the weld metal will immediately coat the tungsten, the arc will start blowing every which way and the welder will have to shut down, break off the dirtied end of the tungsten, grind a new point on the end of the electrode, and put the torch back together before continuing.

Tungsten electrodes are expensive, so **never dip the tungsten into the weld pool, and never touch the tungsten electrode with the filler rod.** Once you learn how much work it is and how long it takes to clean up the mess, it will not be necessary to remind you of this.

One other reason for never dipping the tungsten electrode into the weld

FIGURE 8-1.
A TIG welding machine. Notice the remote control foot pedal which is used to increase or decrease the amperage.

pool is that tiny flaked off pieces of electrode will remain in the weld, and they are much harder to grind smooth than the surrounding metal.

Tungsten electrodes range from .010″ in diameter up to 1/4″ or more, depending on the type and thickness of metal that is welded. Electrodes are made of either pure tungsten or tungsten with a small percentage of thorium, usually 1 to 4%. This small amount of thorium seems to reduce the amount of current needed for welding. Pure tungsten is usually used to weld aluminum, which requires much more heat than other metals. Figure 8-2 shows two TIG torches and their tungsten electrodes.

## GASES

The gases used in TIG welding are usually pure argon or a mixture of argon with helium or some other inert gas. Argon is heavier than helium and conducts heat away from the weld faster. Helium is so light it tends to float away instead of protecting the arc, but helium will conduct electrical current and is less dense, so it

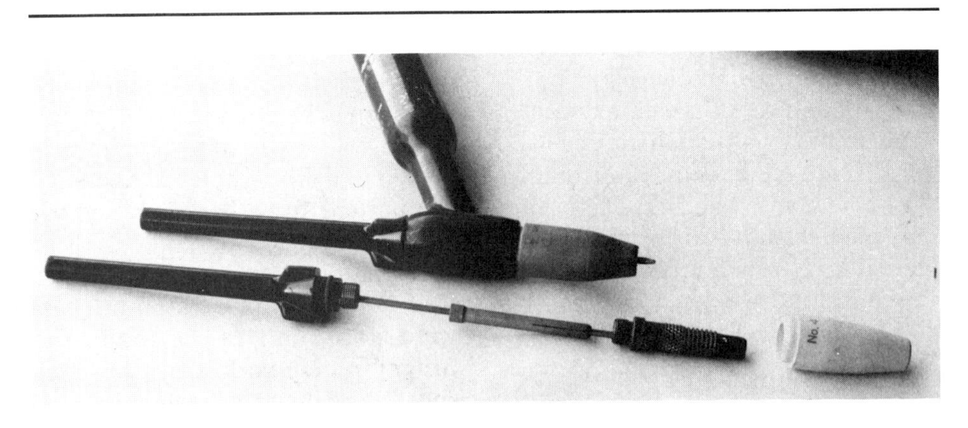

FIGURE 8-2.
A TIG torch assembled for use and another broken down to show its parts.

FIGURE 8-3.
A TIG flowmeter. This instrument is a single stage regulator like those used in oxy-acetylene welding to measure the flow of inert gas to the TIG torch.

Union Carbide Corporation Linde Division

produces a hotter arc. High current welds usually require a compromise in the mixture of argon and helium.

First we will consider pure argon, which is taken from the atmosphere in the same way as oxygen: by compressing air into a liquid and boiling off the gases at their different temperatures.

Argon may come in standard compressed gas cylinders or in liquid form. Liquid argon is stored in thermos-like containers at 235 psi. As the gas is used, this pressure will remain constant because of evaporation from the liquid. The gas must be at least 99.8% pure and must be dry. The other .2% is nitrogen.

Helium is also stored as a compressed gas. Scientists first discovered helium in the sun's spectrum. Later a field of it was found in Kansas. Since Kansas is closer than the sun, most of the world's supply still comes from Kansas.

All cover gases for TIG and MIG processes are measured with a flowmeter (Figure 8-3). This device has a vertical glass column with a slight taper and a tiny pith ball inside. Flowing gas pushes the ball up the column, which is calibrated in cubic-feet-per-minute (fpm) of gas flow. Unless the flowmeter is mounted with the column in a vertical position, the reading will be off.

Preparation of the tungsten electrode is very important. To weld steel the electrode should be ground to a sharp point like a pencil. A properly ground electrode is shown in Figure 8-4. The welding itself should be done with DCSP (direct current, straight polarity). *Straight polarity means*

FIGURE 8-4.
A properly sharpened tungsten electrode. The point should be ground back about three times the diameter of the electrode. A sharpened electrode should be used with DCSP current.

*positive ground, torch negative. Reverse polarity (DCRP) is torch positive and negative ground.* DCSP will produce a stable arc and good penetration with a minimum of porosity. Any time reverse polarity (RP) is used, reduce the amperage by at least one-third of the amount necessary for SP welding. Reverse polarity will burn up the tungsten very rapidly, so a lower amperage must be used.

A tungsten electrode of proper size must be matched with an appropriately sized ceramic cup. See Figure 8-5. Using a cup that is too large for the tungsten electrode will give an uneven distribution of the argon gas, which will cause porosity and other failures. If the cup is too small the lip will burn off.

By now you have learned a few chemical symbols and you probably expect them to be initials or abbreviations for the name of the element: *C*, for instance, is carbon, *N* is nitrogen, and *H* is hydrogen, *He* is helium,

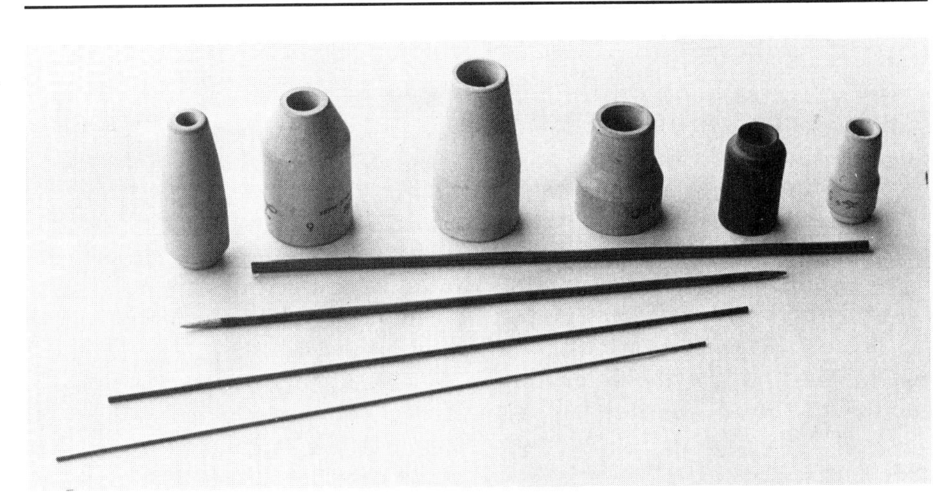

FIGURE 8-5.
TIG tungstens and ceramic cups of different sizes and styles. The correct size tungsten must be used with the correct size cup for proper distribution of the inert gas.

etc. But tungsten is neither *T* nor *Tu*. The chemical symbol for tungsten is *W*. The only, and not very satisfactory, explanation is that it comes from an ore called wolframite. Tungsten is actually the name of a town in Sweden where wolframite was first mined. The only reason for telling you all this is that someday you may be working with European blueprints (which nowadays are seldom blue) or in countries where tungsten is called wolfram.

## REMOTE CONTROL UNIT

TIG machines have a separate small transformer apart from the welding current. This furnishes a low amperage current with extremely high frequency and voltage, which allows the current to leap across a gap just like an automobile spark plug. In this way it is never necessary to touch the electrode to the grounded metal to strike an arc. **Never touch the tungsten with weld metal or rod.** The high-frequency spark jumps from the electrode to the grounded metal to be welded, and the heavy amperage welding current immediately follows it across the path of weakened resistance through the air. To start the arc you simply place your torch nearly horizontal to the plate and depress the foot pedal on the remote control unit (Figure 8-6).

The remote control unit is a necessary accessory for a TIG welder. When the pedal is depressed this unit activates solenoids, which open valves to start the flow of cooling water and argon gas to the torch. These units also control voltage and amperage within preset limits, allowing the operator to feed a hot arc to the cold plate at the beginning of a seam and to gradually cool it as the work warms up. This foot pedal control is very helpful for putting a neat finish on the end of a seam. Each machine is different, so your instructor will have to thoroughly demonstrate the operation of the machine you will use before you can proceed with actual welding. *Basically, you should set the machine low and use the pedal to increase current.*

## PUDDLING

For your first exercise, start the arc, and, beginning at the edge of the plate, melt the base metal to form a pool of liquid metal. Study Figure 8-7 for proper torch angle. Try to keep the pool of molten metal the same size as you advance the torch across the plate. This is like acetylene welding. The big difference is that the TIG

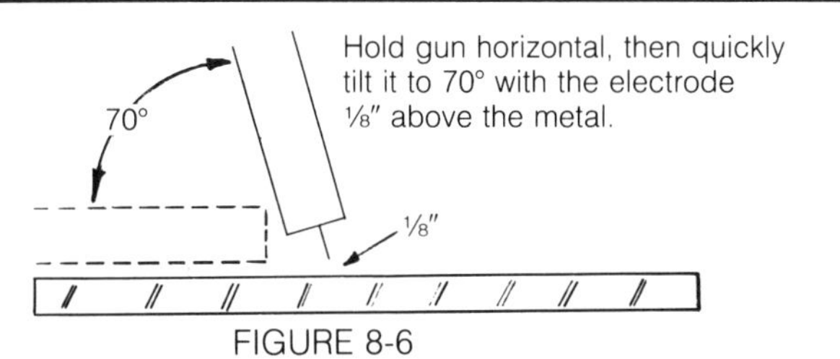


FIGURE 8-6

Striking the arc in TIG welding, flat position. Never touch the tungsten to the plate. The arc will jump across the gap when the remote control pedal is depressed.

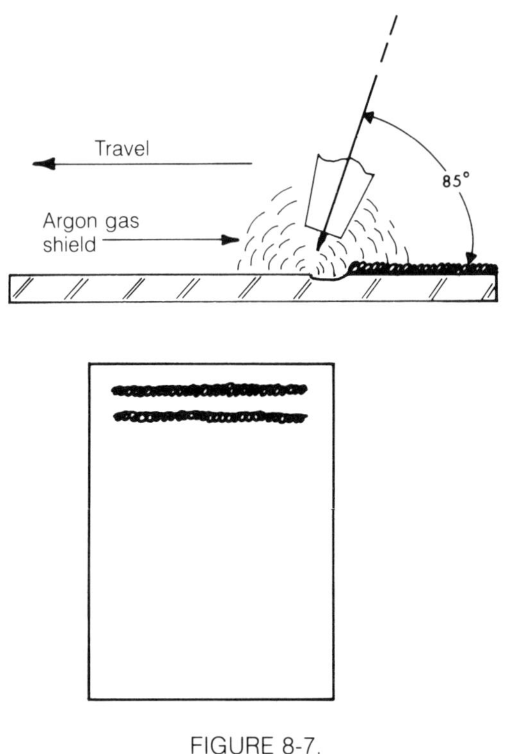


FIGURE 8-7.
Pushing puddles across the plate.

arc is roughly twice as hot as an acetylene flame, so the work melts faster and the seam can be run faster without the heat soaking into the work so deeply. Practice until you can hold a steady arc. Keep your beads about 1/4″ apart and check the back of the plate for penetration. If speed, heat, and torch angle are correct, penetration will be even all the way across the plate.

The TIG arc is not shielded by smoke and the cover gases tend to ionize, as in a neon sign, so it may be necessary for you to change from the normal #10 filter lens to a darker #11 or #12. A TIG arc is very bright, and if the proper shade of darkness is not used, your eyes will tire very quickly—maybe too quickly.

Practice pushing puddles until all the beads are smooth with an even line of penetration on the back of the plate. If the bead tends to sink as your torch progresses across the plate, your amperage is probably too high or your progress across the plate may be too slow. Bulges on the back of the plate should be avoided (Figure 8-8). If your torch is run too fast the beads will appear narrow and not show penetration.

## STRINGER BEADS

***Flat Position.*** After you can push a neat puddle across the plate you should start running beads. To do this you will have to use more heat to

FIGURE 8-8.
Back side of a plate welded in the flat position. Notice the burn line and bulges where the arc nearly burned through the plate.

FIGURE 8-9.
Plate showing size of beads made by different sized tungstens and filler rods.

FIGURE 8-10.
Correct rod and torch angles for TIG welding in the flat position.

melt the filler rod, and your torch travel will be slightly slower. Run single beads across the plate, at least ¼″ apart. Try to keep these beads straight and equal in size. Try different sized tungsten electrodes, making sure to use the proper size cup with each change. Note the effect of different tungsten electrodes and different rod sizes on the bead (Figure 8-9).

Feed the rod pretty much as in acetylene welding, dipping it in the pool as needed. With practice this motion becomes automatic. Correct rod and torch angles are shown in Figure 8-10. After cleaning up the mess a few times you will not have to be reminded to **never touch the tungsten with weld metal or rod.**

Practice this same exercise by running vertical, overhead, and horizontal beads.

***Horizontal Position.*** The angles for torch and filler rod will change for the horizontal position (Figure 8-11).

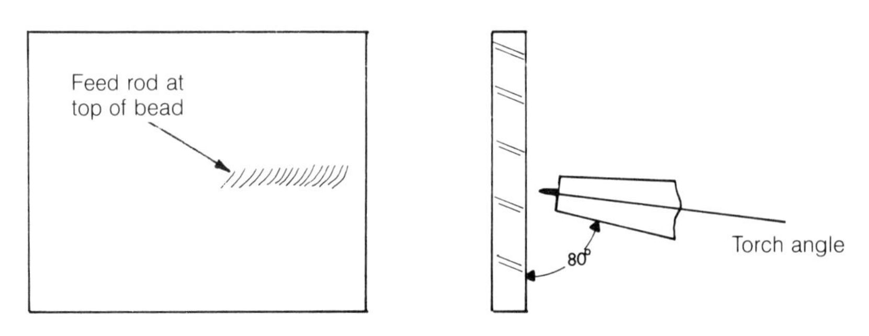


FIGURE 8-11.
Keep the weld bead slightly ahead and at the bottom of the horizontal weld. This forms a small slanted shelf upon which the bead may be built up. The rod should be fed at the top of the bead.

The proper rod and torch angles are necessary to prevent undercutting and rollover. Be careful feeding rod and make sure to feed enough. Again, place these beads ¼″ apart. You will probably have the usual problem of molten metal running downhill as the bead progresses. Only practice and concentration can prevent this natural tendency. Horizontal TIG welding is not difficult, but rod and torch angles are important.

***Vertical Position.*** The vertical position differs in the way rod is fed to the weld puddle. (See Figures 8-12 and 8-13.)

In this position start with a smaller diameter rod, usually about 3⁄32″. With smaller rod a smaller bead will form, and the weld puddle can be controlled more closely. Dip the rod into the weld puddle frequently. Keep the torch pointing at the plate as near 90° as possible. This helps assure gas coverage all around the weld, which is necessary to prevent the oxygen and nitrogen in the air from entering the weld.

Don't try to put on a lot of bead during one pass. It will make a rough looking weld, and it can cause porosity. It saves more time in the long run to make an extra pass or two than to hurry and have to re-do the whole weld. You should learn this position with 1⁄16″, 3⁄32″, and ⅛″ rods, being sure to use the proper size tungsten electrodes and ceramic cups for each rod. Keep the beads ¼″ apart so you can watch the area surrounding the weld as well as the weld itself. The ¼″ will give you room to watch for undercutting and other flaws. This

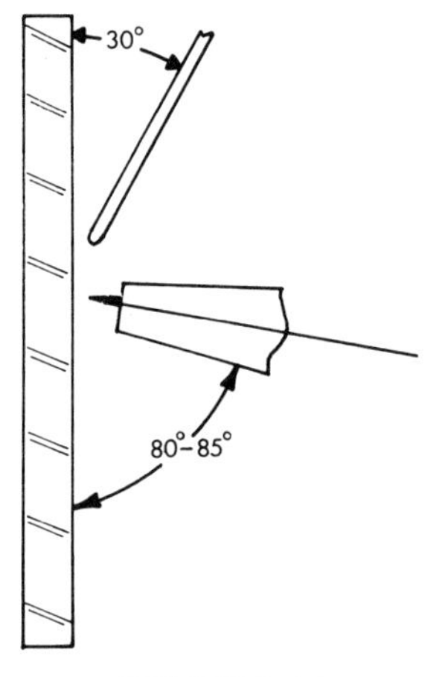


FIGURE 8-12.
Vertical welding. Try to keep these rod and torch angles constant throughout the weld.

FIGURE 8-13.
A welder demonstrating the proper torch and rod angles for vertical TIG welding.

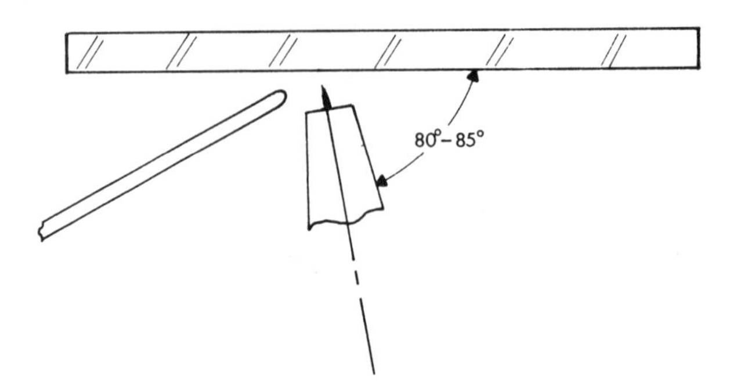


FIGURE 8-14.
Rod and torch angles for overhead welding with the TIG torch.

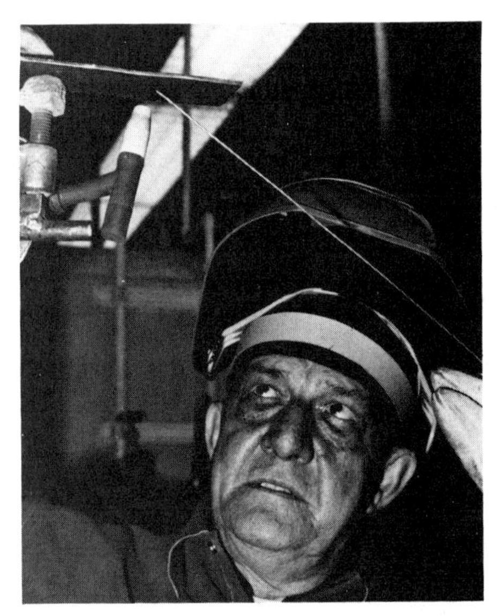

FIGURE 8-15.
The correct rod and torch angles for overhead TIG welding. Get in position but don't strike that arc until the hood is down.

is a certification test position, so all exercises must be practiced to perfection.

***Overhead Position.*** As we never tire of pointing out, except when welding under zero gravity conditions, sparks and molten metal fall *down*. When running overhead beads, all the usual leathers and other safety equipment, plus a liberal ration of common sense, must be employed. Figures 8-14 and 8-15 show rod and torch angles for TIG welding in the overhead position.

This exercise is also a test position, so it must be practiced until you can do it perfectly. Begin with a 3/32″ diameter tungsten electrode and 3/32″ rod. The 3/32″ rod will be large enough for good control in placing weld metal, but it will not be large enough to threaten you with a molten shower. Before beginning any overhead weld the student should stop a moment to consider where the sparks will fall, then figure some way not to be there. Burns are painful and are acquired much more rapidly than they heal.

By now you will have learned that, as was true in acetylene welding, dipping the rod frequently into the weld pool makes a smoother seam. In TIG welding, however, the torch should not be oscillated until using weave passes.

In production welding many things are too large to be turned over for flat welding, so all four welding positions must be mastered.

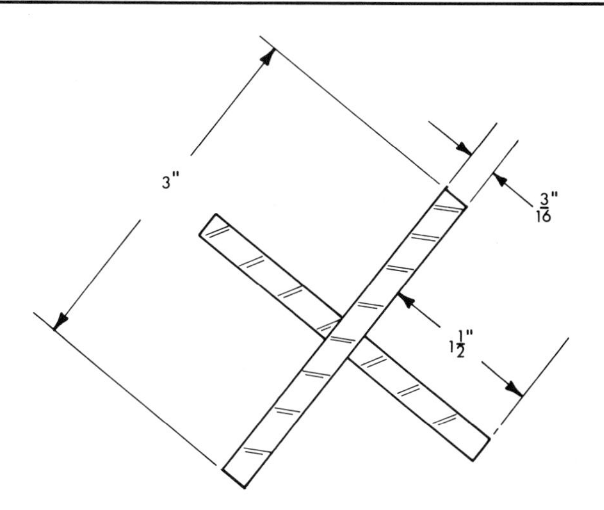


FIGURE 8-16.
Setup of plates for deep Vee or Tee joint practice. An extra plate is placed on the back of the wide plate to save material. This provides two extra corners for practice welding.

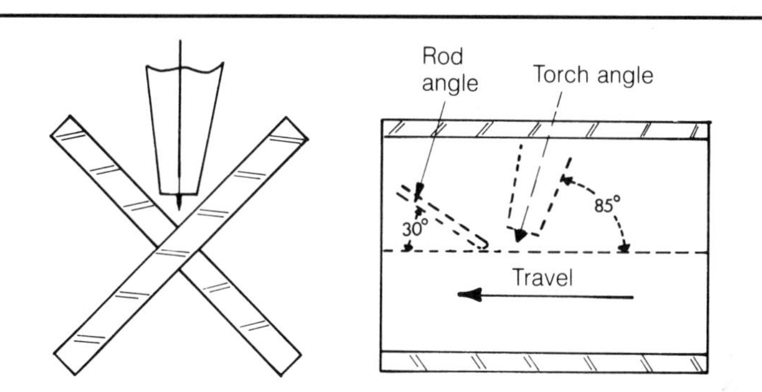


FIGURE 8-17.
Torch and rod angles for deep Vee welds in the flat position.

## DEEP VEE JOINTS

Once you have perfected the art of running stringer beads in the four basic positions you should progress to deep Vee or Tee joints, which must also be done in all four positions. Use three pieces of metal for these joints, cut and assembled as in Figure 8-16.

***Flat Position.*** Vee joints take more heat to weld, since there are now two plates plus the filler rod to be melted. The tungsten electrode will have to be readjusted in the collet until it extends a full 1/4" out of the ceramic cup. If this is not done the lip of the cup will burn off, because the confined space of the Vee reflects heat and blows hot gas back up over the torch. It will take you about two seconds to learn how much heat actually comes back onto even a well-gloved hand.

See Figure 8-17 for torch and rod angles. These angles are only approximate. Frequent dipping of the filler rod is important to prevent undercutting. Not enough dipping will cause undercutting, and too much rod in the root pass will cause a raised weld bead with no penetration. This weld bead should come out flat. Run the root pass in all four corners of the weldment before trying to finish the rest of the weld. This is a better way to learn root passes than to finish each weld independently. It also helps prevent the weld from becoming too distorted by heat from its original shape.

On the second pass, the heat should be directed at the very edge of the root pass (either side), and run the full length of the weld in this manner. Don't try to put on too much weld in one pass, because it will cause porosity and turn the bead face into a convex surface. This must be avoided. Each bead should be flat faced and smooth. After the second bead is applied, the third is run on the opposite edge of the root pass. When the third bead is finished the weld face should be smooth and flat, with no undercutting.

Finish all four grooves of the weldment before running cover passes. These cover passes are put on in about the same manner as those used in acetylene welding. Use a cross pattern, because by now the weld is wide enough to accommodate the added rod. Do not weave any farther than the edges of the weld already applied or there will be undercutting.

Molten metal will not flow up the slanted side of a Vee joint, but it will flow down. If heat is applied to this joint without filler rod an undercut will appear. As the torch moves across the weld face, rod must be applied at the point where the arc contacts base

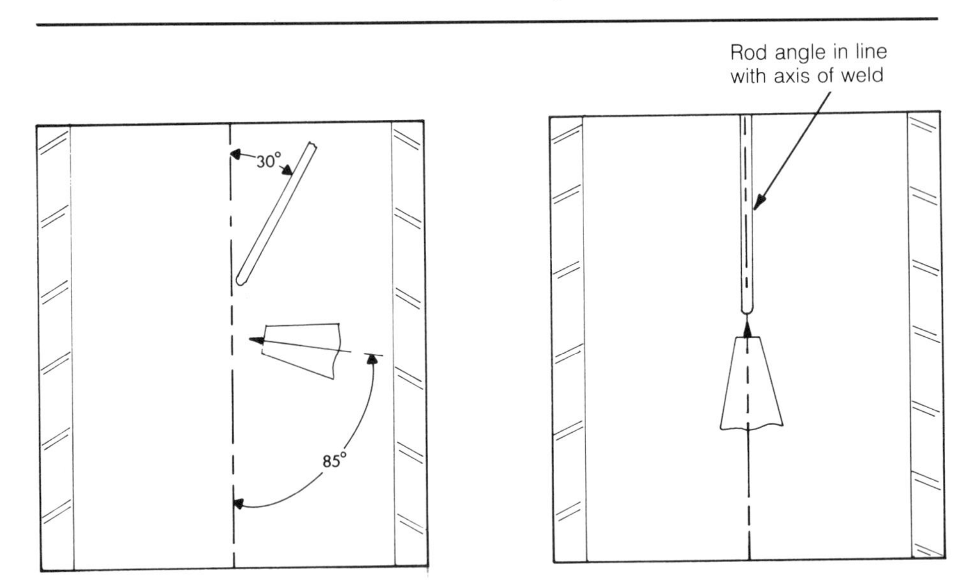


FIGURE 8-18.
Torch and rod angles for deep Vee welds in the vertical position.

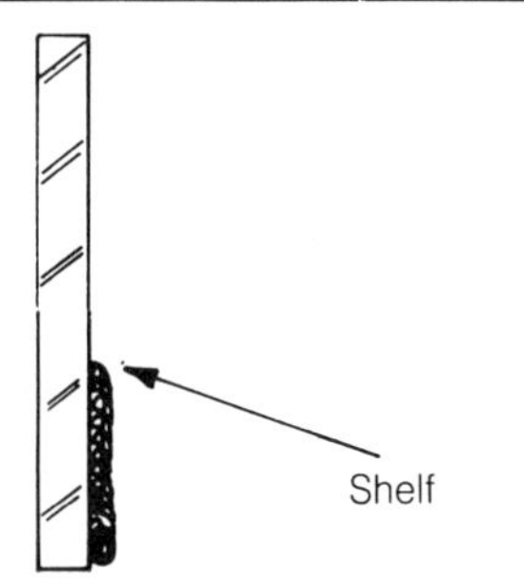


FIGURE 8-19.
A small shelf formed to hold the molten metal.

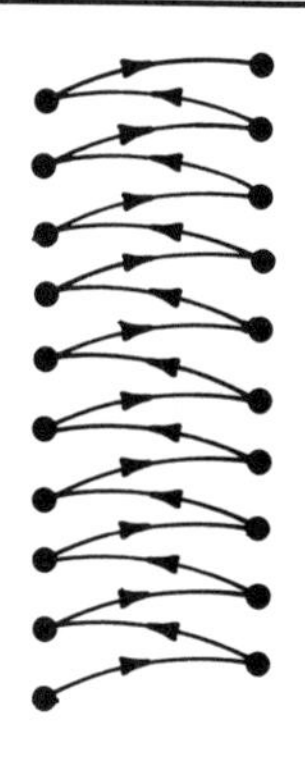

FIGURE 8-20.
Motions of the TIG torch when making a weave pass.

metal. Pause on each side of the seam to make sure the edges of the weld are filling properly. You should weld at least two assemblies in this position to make sure you know what is expected on this joint in the other positions of welding.

***Vertical Position.*** Once you are ready, it's time to move on to the vertical position, which, for TIG welding, is easier than the horizontal position. When starting the vertical weld with a TIG torch, position the plates as in Figure 8-18.

The torch and rod angles noted in Figure 8-18 are only approximate and may have to be changed from time to time to place the weld where it is needed. The distance of the arc from the plate is very important, as well as the placement of the filler rod.

As in the last exercise, apply root passes in all four corners first. All passes in the vertical position should be applied in the same sequence used in the flat position. There should be a minimum of three beads on each corner of the weldment before applying the cover pass. Too much weld at one time will spoil the work. This is the student welder's most common mistake. You must remember that a large amount of weld metal is harder to control than a small amount.

The cover pass is not as hard as it seems. The rod must follow the point of contact with the arc at all times. Whether right- or left-handed, the student will do it the same way. A shelf of metal must be started to keep the weld in position. Begin at one side of the weld and apply enough heat to melt the base metal. Apply filler rod and move the torch across the weld face so a small shelf is formed to hold the molten metal. See Figure 8-19 to see how this is done.

Follow the arc across the weld face with the rod, dipping it as needed. Don't put too much rod in the weld pool or it will tend to run off the weld face. It takes practice to achieve the coordination to make a good cover pass on this weld. Figure 8-20 shows proper torch motions when making a weave pass.

The hardest thing to overcome is the tendency of the arc to keep climbing the slanted plates that form this joint. Undercutting will appear very

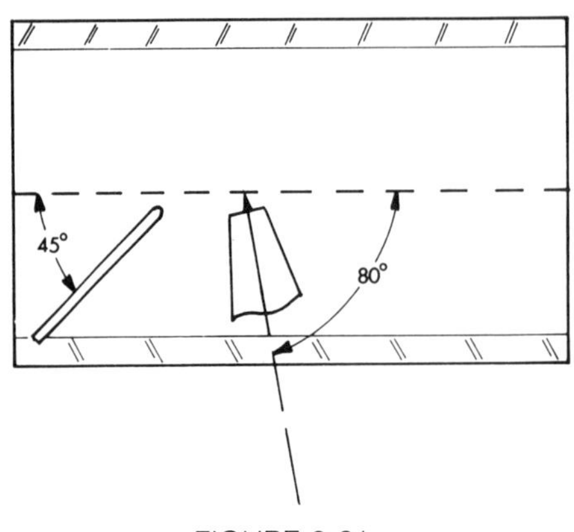


FIGURE 8-21.
Torch and rod angles for overhead deep Vee joints in TIG welding. **Watch the hot ones that may fall!**

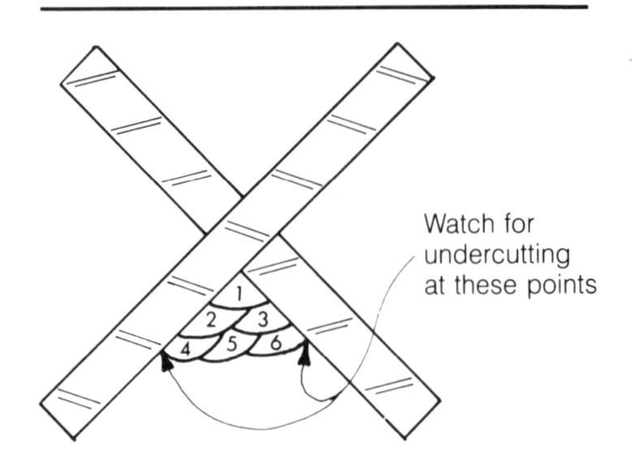

FIGURE 8-22.
Overhead welding sequence if all beads are applied as stringer passes. Watch for undercutting at the edges of the weld on beads 4 and 6.

quickly if the arc is applied to these slanted surfaces. A very definite pause is necessary at the edge of the weld to make sure the weld is taking at the edges. The three stringer beads made before the cover pass will fill the corner of this joint to about the same width as the bead in the 1/4″ grooved plates that will be used for certification.

The real reason for doing these Vee welds is to give you a chance to get used to working the torch in a deep groove. After this exercise, the certification test plates will be much easier to weld correctly. Running a few of these joints seems to shorten the time it takes to prepare for the certification tests.

***Overhead Position.*** All overhead welding is dangerous. Vee joints are no exception. Though the deep Vee may keep sparks from scattering, the operator who gets far enough away to use the Vee as a shield will not be able to see what he or she is doing. This position is also difficult because the torch, hoses, and cables soon become unbearably heavy.

Torch and rod angles are very important when welding the overhead Vee. (See Figure 8-21.) Note that the torch is never changed from the 80° angle in the direction of travel. In this overhead position the gas flow may be slightly lowered. This will depend on the sizes of the tungsten electrode and ceramic cup being used. Distribution of the covering gas is of vital importance, as well as correct heat or amperage and feeding of the filler rod.

As in the preceding exercises, three stringer passes should be run in all four Vees before running cover passes. Finish the entire weld by using stringer passes; or, make at least three stringer passes and one weave pass to cover. Use the sequence in Figure 8-22 if you use the all-stringer-bead method.

If the weave pass is used for the cover pass, pay attention to the edges of the weld where undercutting can appear. Don't try to put on too much weld at once. It's better to run an extra pass than to try to do it all at

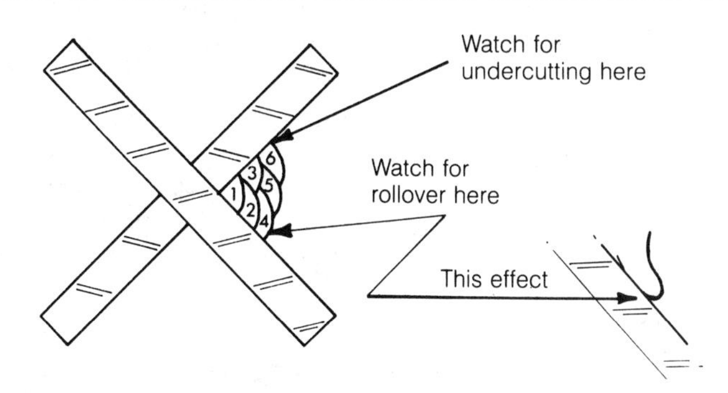


FIGURE 8-23.
Horizontal welding sequence. Watch for rollover at the edge of bead 4 and undercutting at the top edge of bead 6.

once. On the weave passes you will have to concentrate on filling the edges of the weld. Follow the arc across with the filler rod and make sure the molten metal is flowing into the edges as it should. This position is very hard on the arms, and you must resist the temptation to hurry.

This is the second of the three test positions and must be learned perfectly. Take special care with rod and torch angles and the amount of rod being fed into the weld puddle.

***Horizontal Position.*** After learning the basics, welding horizontally on a flat plate was not too difficult. The Vee plate, however, will not be as easy. The root pass is not difficult. It's only a straight stringer bead across the plate, but the passes that follow are harder to perfect than the root pass.

The rod must be applied carefully in the sequence shown in Figure 8-23 to make sure no undercutting appears on the last pass. This is the final position of the certification test, and it must be perfect. With this weld a condition called overlap may occur (Figure 8-24). Overlap usually comes

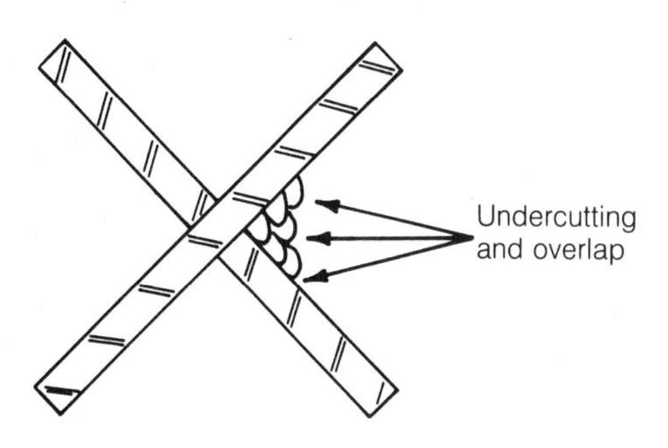


FIGURE 8-24.
A condition called overlap. This is caused by too much heat, feeding too much rod, or holding the wrong torch angle. **Use your foot pedal.** Never set the machine at the point where a perfect weld is obtained with the foot pedal fully depressed. You may need a little more heat.

from feeding the rod at the wrong angle, or from too much heat. Since you can use the foot pedal to control the heat of the arc, you should be able to correct this fault quickly once you have noticed it or your instructor points it out.

Most students think they need the extra heat to place the weld metal exactly where they want it, but they should actually decrease the heat. With this weld, lack of penetration is unlikely, since it is clean, with no slag problems and no way for other foreign matter to enter the weld (unless the welder dips the tungsten). The greatest problem is porosity, which may be caused by lack of heat. When you have learned this position you should practice on grooved plates cut exactly to the size used for the certification test.

## CERTIFICATION TESTS

Before starting practice on the certification plates we should discuss the tests and what is expected of the welder. Also see Appendix D, "Certification."

Certification test welds are tested in a guided-bend test machine. Two strips 1½″ wide will be cut from the test plate across the weld, the edges will be ground smooth, and the test strip will be placed in this machine. One piece will be bent against the root of the weld and the other will be bent against the face. No fracture or sign of separation is permitted in either direction.

Test plates are approximately 8″ long, 5″ wide, and beveled to a maximum of 30°. A backup plate may be used, or the weld may be made with an open root. The backup plate can be used for control of penetration, but it has to be removed before the actual testing or bending of the welded plates.

## PRACTICING FOR THE CERTIFICATION TESTS

***Flat Position.*** If the certification tests are taken with an open root, the root cannot be wider than ⅛″ or nar-

FIGURE 8-25.
Two weld samples that failed certification tests. Both samples show a lack of fusion between the weld metal and the base metal.

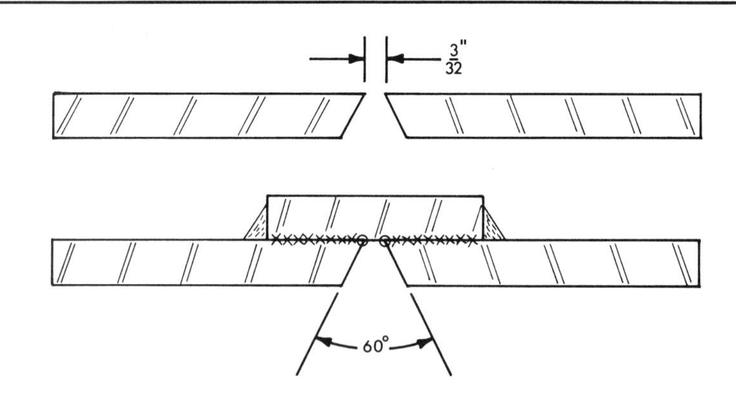


FIGURE 8-26.
Two setups for welding TIG certification plates. When a backup plate is used, it must be removed by grinding and polishing before the weld is tested.

rower than 1⁄16″. A 3⁄32″ welding rod makes a good spacer for measuring this root opening.

Study Figure 8-26 and set up the plates as described. It's best to weld at least one set of plates in the flat position and then bend the samples to be sure you're getting proper penetration. Weld one set of plates with a backup plate and one with an open root to find which way you can do it best. Torch and rod angles are the same as for Vee joints, but close watch must be kept on the root pass. Too much penetration can blow a hole.

***Keyhole.*** The keyhole described in the oxy-acetylene welding chapter is again needed, and it must be kept the same size from one end of the weld to the other. To do this, the remote control unit must be used constantly.

If the keyhole becomes too large, the remote control unit must be adjusted to lower the amperage. It's the same as slowing down a speeding car by releasing the pressure of your foot on the accelerator: if the pressure is released on the remote control unit, the amperage to the arc will be lowered. Proper use of the remote control unit will allow you to maintain a perfect balance of heat to the weld zone. The size of the filler rod being used will affect the use of the remote control unit. Recall that to melt any of the mild steels, both the rod and the base metal must be brought from room temperature to about 3,000°F, requiring constant use of the remote control unit.

***Backup Plates.*** It is a little easier to make your first weld with the help of a backup plate, but this backup plate also makes the weldment harder to prepare for testing, and considerably more heat is required to achieve penetration on the root pass. Also, more filler rod can be put into the root pass. When using this backup plate, however, make sure that the two plates are being welded to the backup plate simultaneously. Too much filler rod will cool the weld zone and cause a lack of penetration, which will show up when the plates are cut and bent in the testing machine.

Once you have decided whether or not to use a backup plate on the test welds, you should not change your mind. It will take that much longer to achieve certification. Remember, you must get 100% penetration for these welds to withstand the guided-bend test machine.

***Vertical Position.*** When the test plates you have done in the flat position will bend without fracture or yield, you should begin practicing in the vertical position. For this position, assemble the plates as you did in the last exercise. Careful assembly of these plates is important.

In welding, a correct fit is half the job, and a weld that fits together poorly will be difficult to finish properly. In the vertical position it can be *more* than half the job. Once certified and working, a good welder will not hesitate to do an assembler a favor once in a while, and cover up the other's sloppiness with some "gap rod," but practicing for certification is no time to set up sloppy joints.

Don't try to deposit too much metal in the root pass of a vertical weld, because this will make the face rough and difficult to finish. Each pass must be as smooth as possible, otherwise it will be impossible to smooth out the irregularities with the cover pass. On the first three passes it is not necessary to move the torch in any way except for the steady travel upward. Again, finish all root and stringer beads before running the cover pass. The cover pass is the most difficult, and requires a steady hand and a close watch on the edge of the weld to prevent undercutting.

Again, the rod should follow the arc and be in contact with the plate at the point where the heat is applied. There should be a definite pause at each edge of the weld during the weaving pattern. Too short a pause: undercutting. Too long a pause: overlap. This position requires much practice. You should pay careful attention to the way you use the torch and the way you dip the filler rod.

The way you dip the rod is as important as proper use of the torch. Your coordination will improve with practice. Practice these welds until your samples can be cut and bent consistently with no sign of fracture or yield. Once you have perfected this position you should move on to the overhead position, at which point we must return to a discussion of safety.

***Overhead Position.*** **Molten metal falls downward.** The drops of metal that fall down on the operator welding in the overhead position, or any other position, have a temperature of about 3000°F. They burn *quickly* and they burn *deeply*. A big gob will go right through your leathers, so *watch it*—**no pants with cuffs, no short pant legs, no open boot tops.** If you carry matches, try not to set them afire. This may seem repetitive, but consider which part of the human anatomy gets burned if matches are carried in a hip pocket. That nameless portion of the human body is *yours!*

Setups for the overhead position are shown in Figure 8-27.

You may find full penetration difficult while welding overhead with TIG. Here, the gas does not push as hard as the gas used in acetylene welding. Penetration is attained by feeding the rod correctly. If proper heat is applied the weld should be level with the top of the plates, and not sunken. A slight back and forth movement of the torch will help achieve penetration.

Don't try to push the root pass

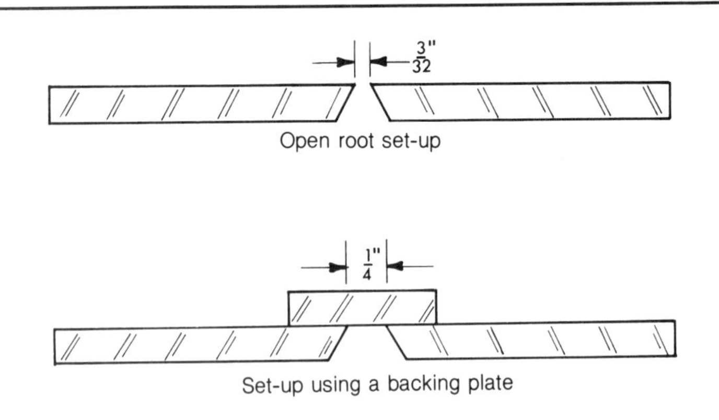


FIGURE 8-27.
Setup for overhead certification. Backup plate must be removed and the root pass ground and polished before cutting and bending.

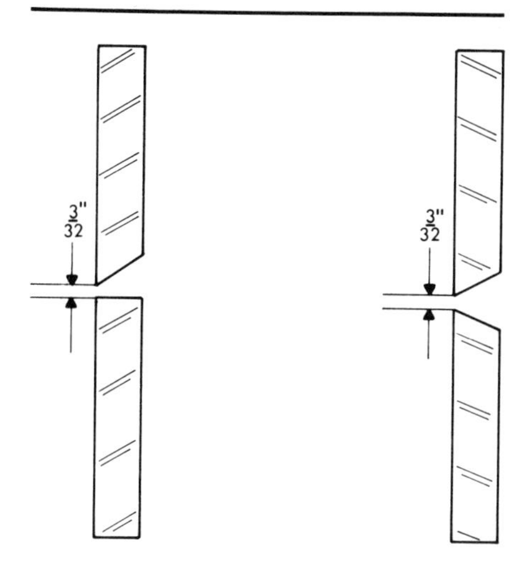


FIGURE 8-28.
Plate setups for horizontal certification tests.

through by increasing the gas flow. This can only cause porosity. Clean the weld after each pass. TIG is cleaner than arc welding, but cleaning after each pass is still necessary. Any dirt in the base metal or impurities in the welding rod will float to the surface of the seam and appear as slag, which must be knocked off with a wire brush or it will be imbedded in the weld.

The final, or cover, pass may be applied as a series of stringer passes or as a weave pass. The weave pass is more difficult because it tries to hang in bulges instead of laying smooth. The easiest way to finish this weld is with a series of stringer passes. As long as the weld surface is at least 1/16" and not more than 1/8" above the plate, the weld will pass the appearance test.

Practice these welds until your samples will hold together consecutively in the guided-bend test machine. Once your welds can stand this test, you should proceed to the horizontal weld.

***Horizontal Position.*** The horizontal position is normally easy to learn, but not so with TIG welding. Both plates may be beveled, or just the top plate (see Figure 8-28).

Penetration is not easy to achieve when only one plate is beveled and a backup plate used, and the greatest problem will be the finish or cover passes. The final bead is very difficult to apply without undercutting. It takes perfect control of heat and proper manipulation of the filler rod to prevent this. The horizontal position must be practiced until your samples consistently pass the guided-bend test.

If you have practiced properly you should now be ready for the certification test. Practice makes perfect, but more importantly, practice makes confidence. Too often, students fail the certification tests not from lack of ability but from lack of confidence. This can be remedied by practicing on plates the same size as those actually used for the certification test. This will help you face the moment of truth as if it were just another practice plate.

## OTHER WELDED JOINTS

By the time you have passed your certification tests you should know most of the basics of welding. You should still practice lap joints in the four basic positions (shown in Figure 8-29), but by now they will not be difficult.

## WELDING NONFERROUS METALS

Nonferrous metals weld much differently than the iron and steel welded so far. TIG welding was first

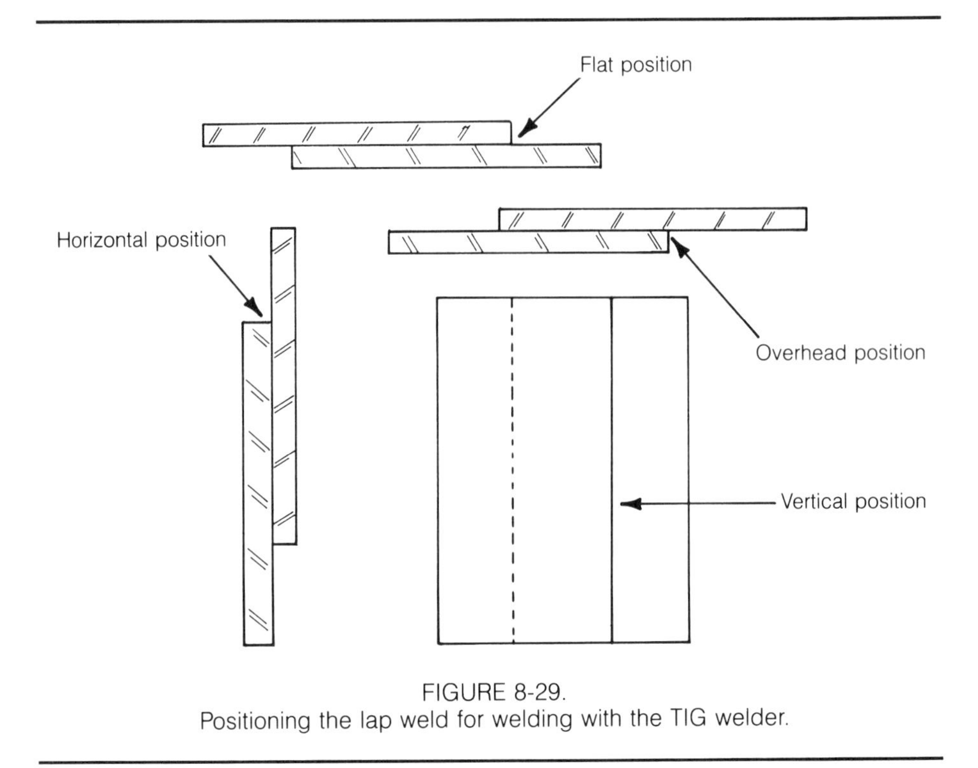


FIGURE 8-29.
Positioning the lap weld for welding with the TIG welder.

developed in World War II to weld magnesium airplane wheels. Nowadays the process is mostly used to weld aluminum. The cover gas is usually argon, with a little helium mixed in. Stainless steels (which contain a lot of copper and nickel) are also welded with TIG.

Stainless steels are poor conductors of heat and can be welded with surprisingly low amperages. Aluminum carries away the heat much faster than steel and requires a very high amperage. Once you are certified in steel, the other metals seem to be much easier to learn.

***Aluminum.*** Aluminum can be dangerous to weld. Since molten aluminum is poisonous, **aluminum burns need immediate attention by a doctor.** Aluminum melts at a lower temperature than steel or stainless. It does not glow or change color, and if the welder does not recognize the sudden "wet" look of the weld pool, he or she may be greeted by an unexpected shower of molten metal, especially when welding overhead.

Since aluminum conducts heat away from the arc zone more rapidly than iron, the unsuspecting student may find a surprisingly large puddle of metal suddenly plopping down. As in learning any other weld, practice in the downhand or flat position first. Aluminum, especially, must be cleaned after each pass. You should learn to weld all types of joints and positions.

***Stainless Steel.*** Stainless steels should be fluxed on the back side of the joint to be welded. Flux the rod, too, for a clean weld, and when finishing, hold the cold torch over the metal with the gas running until the bead stops glowing. This keeps air from the hot stainless steel and prevents it from burning.

When TIG welding stainless it is important to match the metal with the filler rod. The preferred cover gas for thin gauge metal is argon, because it demands less heat than helium. However, a faster rate of travel may be obtained by using helium. DCSP is recommended, but AC can be used on light gauge stainless to decrease heat input.

The "stainless" quality of this steel is due to chromium and nickel. Excessively high temperatures will burn these metals out and the steel is no longer stainless. Since stainless steel is a poor thermal conductor (40-50% less than plain carbon steel) heat stays in the weld longer than you expect. It helps to remove this heat with a copper backing plate. **Never use a brass chill block around stainless. Brass is an alloy of copper and zinc, and any zinc contamination of stainless steel will cause cracking.** It is also a good idea to use jigs and fixtures made of materials that are good thermal conductors.

Stainless expands more than carbon steel. This expansion brings the danger of cracking. To prevent it, preheat the metal 300°-500°F and avoid welding continually in a small area. Heat buildup can alter the microstructure. *Microstructure* refers to the microscopic crystalline form into which a melted metal hardens. Tempering and annealing processes are all used to change a metal's microstructure.

When welding stainless with acetylene, a slightly reducing flame sometimes creates a smoother weld, but this can also weaken the bond. A reducing (carburizing) flame is not as hot as a neutral flame. It can contaminate the weld with both carbon and hydrogen.

Types of current vary with these metals. Aluminum is welded with ACHF (Alternating Current High Frequency). Steels are usually welded with DC, either straight or reverse polarity. When using reverse polarity, lower the amperage to prevent burning up the tungsten electrode. These metals are more expensive than plain carbon steel. Scrap metal for practice is harder to find.

***Copper and Cuprous Alloys.*** Cuprous alloys are copper based, such as brass (copper and zinc), bronze (copper and tin), and other alloys containing copper plus nickel, aluminum, lead, phosphorus, silicon, or beryllium. Like pure deoxidized copper, most of these can be TIG welded. As the amount of tin, lead, or zinc increases, TIG weldability decreases, because these elements boil at arc temperatures. Such alloys should be oxy-acetylene welded.

Beryllium-copper is hard and often used to make nonsparking pliers, screwdrivers, and other safe tools for handling explosives. **Beryllium is poison!** Dust from grinding or fumes from welding can do as much damage as breathing plutonium. Unlike atomic wastes, beryllium has no half life. Beryllium lasts forever!

Copper readily conducts both electricity and heat. This means the arc must be hotter to produce a puddle, because heat is pulled away into the surrounding metal. Current settings are 50% to 75% higher than for the same thickness of aluminum. DCSP is most often used, although AC is not uncommon. Use a mix of argon and helium. Molten copper's toxic (poisonous) fumes require maximum ventilation.

Copper is *hot short*. This means it becomes especially unstable just under its melting point. The slightest weight or shock when the metal becomes a medium cherry red causes cracks or collapse. Copper should be backed, either by a copper backing bar, if the weld will not penetrate into the backing, or by oxidized stainless steel. Stainless steel is oxidized by heating it in the presence of air. Stainless has a thermal expansion rate similar to copper's but low thermal conductivity, and does not drain heat away so rapidly.

There are many job opportunities in TIG welding. Its greatest use is probably in the aerospace industry. The weld is clean and gives excellent penetration, because the heat is constantly controlled by the foot pedal. All industries dealing with aluminum require TIG welding. Large factories and most industries with a maintenance department will have use for TIG welding. Small welding shops usually have TIG welders for repair and custom work. A good TIG welder will probably never have trouble finding a job.

# 9 CAREERS IN WELDING

Brute strength is not important. Good eyesight is necessary. Since flame and metal colors are important, it also helps not to be color-blind. Careers can be built on many levels of education and skill.

Those who settle for routine jobs will discover that it is impossible to stand still. The world moves. They lag behind until finally, middle-aged and tired, they must face the bitter truth that hard work is not enough. Only an inquiring mind and constant study can keep one's skills up-to-date and in demand.

To make it big, learn to spell and to write simple, clear English so the bosses will read your reports. Study drafting, blueprint reading, shop, physics, general science, chemistry, and mathematics.

## PREPARATION

*Practical Welding* in high school or vocational school prepares you for entry into the trade. To stay in the trade demands constant reading to keep up with new developments and to upgrade skills.

Factory owners have learned that robots never go on strike. Robots are willing to work nonstop all three shifts. They demand neither coffee breaks nor a parking lot. In the United States mufflers, washing machines, gas tanks, and countless other items are already produced by automatic welding machines. Trains run on automatically welded rails. Petroleum pipeline and storage tank welding are increasingly automated.

Drilling rigs and refineries, which are one huge weld, are still put together by skilled professionals, but the machines are gaining. The future in production welding is not in welding itself. The future belongs to the person who knows how to program and supervise the robot.

Far more skilled welders work on high-technology research and development prototypes in laboratories and research establishments, putting together one-of-a-kind units and stretching their craft to its limits. These are the fun jobs. Part of the fun is the good pay.

Pay depends on knowledge and skill. The best-paying jobs in welding naturally require the most education. Rule-of-thumb is that every year of school adds another $15,000 to your lifetime earnings. To get into most engineering colleges you'll need:

3 units of English composition and literature

2 units of social studies—American history, civics, etc.

3 units math—geometry, algebra, trigonometry

3 units science—chemistry and physics

5 units additional work in the above and other subjects

A foreign language is also sometimes required. *Unit* is one full year's credit for a subject.

## GETTING YOUR FOOT IN THE DOOR

High schools, industrial schools, and vocational-technical schools all offer courses in welding. Occasionally, in booming times of full employment, one can get a job with very little actual experience or skill. Only an optimist counts on it.

The more experience, the easier it is to get a job. Those who can count on their parents, or have private means, can afford the more expensive schools. The majority must find someone willing to train and pay them.

Most states run apprenticeship programs. Check the phone book and the library for information.

The Navy Department, working through the federal Civil Service System, offers an apprentice program for welders. Pay and working conditions are excellent, but there are more applicants than openings so apply ahead of time. Years ahead. See the phone book for your local Civil Service Commission office.

If you join the navy and go to welder's school, you'll be rated as a boilermaker. After the enlistment's up a boilermaker can find a good job. Recruiters are usually near the Post Office. Other branches of the armed services offer technical schools. The Army Corps of Engineers builds bridges, pipelines, radio towers, and buildings.

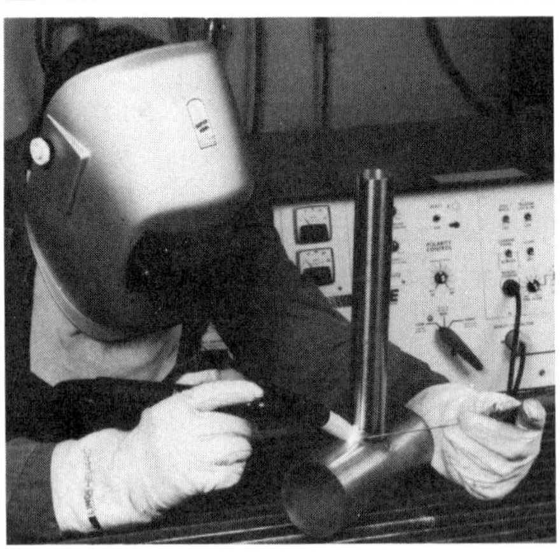

FIGURE 9-1
A TIG welder welding pipe in a Tee joint.

Union Carbide Corporation Linde Division

## JOBS FOR WELDERS

**Tacker:** This person tack welds assemblies to hold them together long enough for finish welding. This is usually the first job you'll get when fresh out of school. Low pay and low skill. Do it well and you'll advance.

**Welding operator:** Low skill, low pay. Runs a machine to perform repetitive tasks.

**Welder:** Pay depends on skill. Steamfitting, shipbuilding, pipeline industries all need welders. Good job for now but don't stop here or you'll be automated out. Keep on learning.

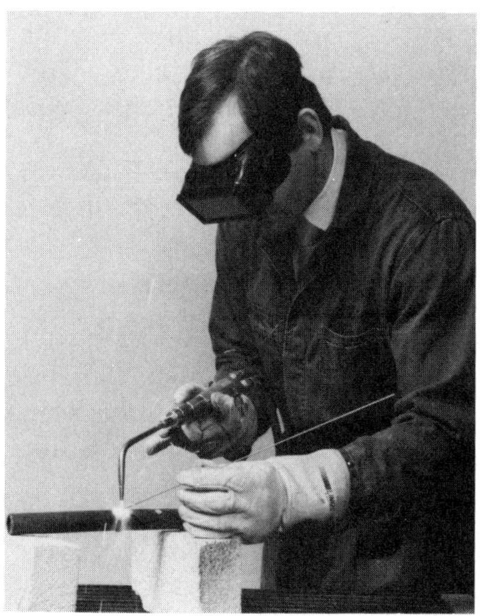

FIGURE 9-2
Oxy-acetylene welder.
Union Carbide Corporation Linde Division

**Auto/truck body repair technician:** Knows welding but must also be skilled in sheet metal work, hand forming, lead burning, and plastic filler use. Average pay.

**Combination welder:** A catchall term that can mean different things in different shops. As in other branches of welding, pay depends on skill.

**Welder-fitter:** Knowledge of other crafts like machining, sheet metal work, structural iron, or shipfitting is required. Doesn't just weld together somebody else's job. The welder-fitter also reads the prints, cuts the metal, and prepares the work.

**Specialist welder:** Has welded long enough to develop a special skill in some hard-to-weld metal and becomes known as the expert. Expertise pays well.

**Maintenance welder:** Usually an older welder with much and varied experience, capable of any kind of welding. Pay is good but not spectacular. This can be a very steady job.

**Welding technician:** Originally, these people worked in labs but now they're out in the shops, using gauges and instruments to collect data and run tests that used to demand an engineer. This job can lead to an engineering degree, or one's own business.

**Supervisor:** Supervisors and foremen have served their time welding. They also know how to get along with people, and how to handle the paperwork involved in running a shop and crew.

**Underwater welder:** These are the test pilots of welding. First, learn to weld. Then learn to dive. *Then* learn to weld underwater. Very high pay for *very* dangerous work—often around offshore drilling rigs. More information from Admissions Director, Commercial Diving Center, 272 South Fries Avenue, Wilmington, CA 90744.

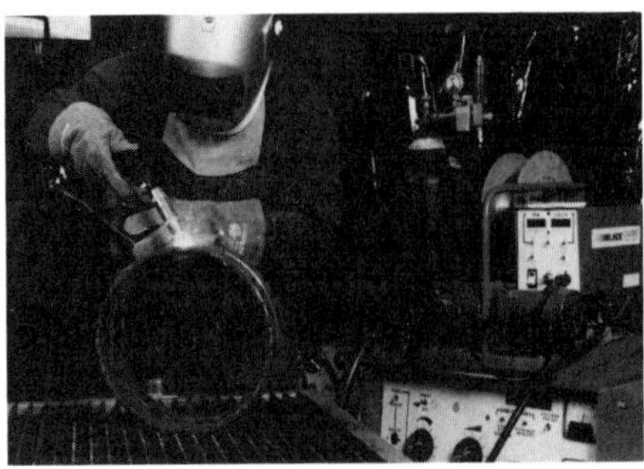

FIGURE 9-3
MIG welder and MIG welding equipment.
Union Carbide Corporation Linde Division

**Welding engineer:** *Development* engineers are the inventors who create new products using tools and techniques that have never been tried before. *Research* engineers create the tools and techniques. A welding engineer must know metallurgy; mechanical, civil, industrial, and electrical engineering; plus all stress calculations and testing techniques.

**Job shop manager:** Usually in business for themselves. Skill in welding is not enough. It takes accounting, record keeping, tax law, and knowledge of the market. Businesses seldom stand still. They grow or go broke.

**Inspector:** These people control quality of the work. They sometimes operate testing machines; read blueprints and weld symbols; can tell a good weld from a bad one. Know codes applicable to the job. Constantly deal with people who wish they'd go away.

**Planner and estimator:** Reads prints, makes up bills of materials, calculates labor and material to do a job. Determines how much to bid if the company is to make a profit and survive.

**Educator:** Instructors, teachers, and professors have usually served their time in the trade, then gone on to qualify in teaching skills. May also write reference and textbooks. Often review technical manuscripts.

**Technical writer/editor:** Knows the trade and how to explain things on paper. Constantly reads to keep up with new developments.

**Sales and service representative:** Must know welding and welding equipment and be able to get along with people. Plenty of money. Hard work and constant travel.

# GLOSSARY

**AC** or **alternating current:** A type of electric current used in households and industry. The most common type of AC oscillates at 60 cycles/second, which means that it reverses its direction of flow 120 times per second.

**Acetylene gas:** A gas whose molecule contains two carbon atoms and two hydrogen atoms ($C_2H_2$). Universally used for welding and cutting, it produces a very hot flame (6000°F) when burned with pure oxygen.

**Alloy:** An intimate mixture of two or more metals, such as copper and zinc, which makes brass.

**Amperage:** The portion of electricity that causes heat in the electric arc, commonly called current; the amount of current flow, as opposed to voltage, which is pressure.

**Annealing:** The softening of metals by controlled heating and cooling.

**Anode:** The positive pole in any electrolytic solution. Theoretically, electron flow is from negative to positive.

**Arc:** The flow of electricity across an open gap. The high resistance of the gap produces heat which, in welding, is used to melt both rod and weld metal.

**Arc blow:** Deviation of the arc from its normal path, caused by magnetic fields being distorted by the shape of the work piece. Arc blow happens only with direct current.

**Arc voltage:** Voltage is the pressure necessary to push the current across the gaseous gap of an arc. The voltage required for welding will vary according to the size of the electrode and type of current being used.

**Arc welding:** Use of an electric arc to deposit filler metal and fuse two or more pieces of metal together.

**Austenite:** Iron with carbon in solid solution. Austenitic stainless steels are non-magnetic.

**Automatic welding:** Welding done by machine without manual handling of the arc.

**Axis of the weld:** The imaginary center of any weld, parallel to the length of the seam.

**Backfire:** A loud banging noise which is usually caused by too low a gas pressure for the tip size being used.

**Backhand welding:** Welding usually done with either acetylene or heliarc (TIG) torch in which the weld is run in the opposite direction toward which the torch is pointed.

**Backing plate:** A plate used on the back of a weld, usually a groove weld, to aid in control of penetration. Under some AWS codes these plates are used in certification.

**Back-seating valve:** A special valve used on high pressure cylinders (such as those used to store oxygen) to prevent leakage of gas from around the valve stem while the valve is open. It should always be fully open or fully closed—never part way.

**Base metal:** Any metal to be welded, brazed, soldered, or cut. The same thing as parent metal.

**Basket weave:** Term for a way of moving the rod in a smooth curve while arc welding.

**Bead:** A row or seam of metal where two pieces have been melted or welded together. Often called *seam* or *weld pass.*

**Bevel:** The angle applied to the edge of a plate to be welded. Bevels may be ground, machined, or flame cut.

**Blowpipe:** Originally a mouth-powered jet of air blown through a charcoal or alcohol flame. Now a common name for an acetylene torch.

**Blowtorch:** A heat source used to heat soldering coppers. Uses common white gasoline for fuel.

**Body:** The main structural part of a regulator or acetylene torch.

**Bond:** The point at the edge of a weld where weld metal meets base metal.

**Box weave:** An alternate method of moving the rod while arc welding. In box weaving the rod moves in straight lines and turns sharp corners.

**Boyle's Law:** Boyle's Law explains the relationship between density, pressure, and temperature of compressed gases. Compressed gases become hotter. Lowering pressure lowers temperature—which is why refrigerators work.

**Brazing:** Joining of two pieces of metal using a brass alloy filler rod which melts at a lower temperature (800″F-plus) than the parent metals. A flux is used to clean corrosion from the surfaces and permit "wetting." The pieces are bonded by capillary action.

**Buildup:** The height of the weld seam above the surface of the weldment.

**Burning:** Violent combination of oxygen with any substance to produce heat. Also flame cutting.

**Bursting disc device:** A specially made safety device on an oxygen cylinder or other high pressure cylinder.

**Butt joint:** Two pieces of metal joined edge to edge on the same plane.

**Calcium hydroxide:** CaOH. Slaked lime; by-product of acetylene gas manufacture.

**Cap:** Another term for *cover pass.*

**Capillary action:** The ability of a metal to move into small spaces by a wetting action. This is especially notable in soldering processes.

**Carbon:** A chemical element, symbol C, valence 4, atomic weight 14.

The principal heat source in acetylene and other gases. Small amounts of carbon added to iron turn it into steel. Used for molds to hold weld metal, arc weld electrodes, motor and generator brushes. Charcoal and coke are almost pure carbon. Absolutely pure carbon crystallized under extreme pressure becomes a diamond.

**Carbon dioxide:** $CO_2$. Used in fire extinguishers, and as a cover gas for MIG welding iron and steel.

**Carburizing:** A flame burning with an excess of fuel. Also called a reducing flame.

**Cathode:** The negative contact in an electrolytic solution.

**Centigrade (Celsius):** The metric system of measuring temperature. Water freezes at 0″C and boils at 100°C.

**Ceramic cup:** A specially made sleeve used on the TIG welder to direct the inert gas into the weld zone. Covers the tungsten at the tip of the torch, and comes in different sizes to accommodate the size tip being used.

**Chamfering:** The cutting of one edge of a plate in preparation for welding. The cut does not penetrate to the opposite side completely and will leave a flat portion at the bottom of the cut.

**Collet:** The portion of the TIG torch that holds the tungsten electrode in position.

**Concave weld:** A weld whose center is lower than its edges.

**Cone:** The innermost portion of a neutral welding flame.

**Continuous weld:** A weld, regardless of length, which is completed in one operation without any break or interruption.

**Convex weld:** A weld whose face is higher than the surface of the plates upon which it is applied.

**Core:** A term used to indicate the center of a welding rod, or, in the case of wire welding or soldering, the center of the wire which will hold the flux that is used.

**Corner joint:** A joint formed by placing two plates edge to edge at a 90° angle to one another.

**Coupling distance:** The distance to which an oxy-MAPP flame cone must be backed off from the weld pool. This distance is considerably farther than with oxy-acetylene.

**Covered electrode:** A metal or alloy rod used in arc welding which is coated with materials to aid in control of the arc and placing of the weld. As the rod burns, the coating forms a gas to prevent air from contacting the weld pool.

**Cover pass:** The final pass to cover a weld. It is usually made with a weave pass pattern.

**Cracking:** Opening a compressed gas cylinder valve slightly to blow dust and dirt from the fittings. This prevents damage to regulators and other parts of the welding system.

**Crater:** The depression at the end of a weld, usually more pronounced at the end of an arc weld bead.

**CRES:** Corrosion RESisting (stainless) steel.

**Crown:** The convex surface of a finished weld.

**CRS:** Cold rolled steel.

**Cutting flame:** Flame used to cut ferrous metals, obtained by use of a specially built torch. This flame usually consists of 4 to 6 pre-heater jets and one larger orifice to emit a stream of pure oxygen once the metal is hot enough to burn.

**Cylinder:** A container used for storing compressed gases.

**DC or Direct current:** A type of electric current which travels in one direction only. In welding it may be made to flow either way. See *Polarity.*

**Dipping (the rod):** The pattern used in acetylene or TIG welding for placing the filler rod into the weld zone.

**Discoloration:** The change of color in metals after welding or applying heat. This is very noticeable in the welding of stainless steels.

**Distortion:** Warping or expansion of metals caused by the heat of welding.

**Downbead:** In the vertical position, a bead of weld whose progression starts at the top of a plate and goes downward. Produces very little penetration of the rod metal into the base metal. Used a great deal on sheet metal welds.

**Downhand welding:** Welding in the flat position with the rod and torch pointing downward.

**Edge weld** or **joint:** A joint formed when two pieces of metal are laid flat against each other with at least two of the edges even with each other.

**Electrode:** That portion of the welding machine from which the arc jumps to the weld pool.

**Elongation:** The amount a weld will stretch before it begins to break or yield to stress.

**Face of the weld:** The exposed surface of any weld.

**Fahrenheit:** The English scale for measuring temperature. Zero degrees is the lowest temperature which Gabriel Fahrenheit could measure in a mixture of ice and salt one particular day in Gdansk during the winter of 1809. In this scale, water freezes at 32°F and boils at 212°F, at sea level. The United States is one of the few countries in the world which still uses this arbitrary and antiquated system.

**Fillet weld:** A weld formed at the intersection of two plates placed at a 90° angle to one another to form a "Tee" joint.

**Filler rod:** The rod that is melted to form the weld proper, regardless of the type of metal being welded.

**Firebrick:** A specially made brick that is used to insulate a furnace or a table used for acetylene welding. The weldments are usually placed atop the brick in order to be welded.

**Flame cutting:** See *Cutting flame.*

**Flash:** The impact of ultraviolet radiation against the human eye. Flash is very painful and can cause blindness.

**Flat position:** Any welding that is done with the work lying flat on a table. This position is also called the downhand position.

**Fluffing away:** The burning away of the edge of the top plate on a lap weld.

**Flux:** An agent used in welding or soldering to clean the weldment surfaces. Many fluxes are metallic salts.

**Forehand welding:** Welding in which the seam runs the way the torch is pointing.

**Free bend test:** The process of bending a specimen without the use of guides.

**Fuse plug:** A safety device that melts at low temperature to release gas from a cylinder, thus substituting fire for explosion.

**Fusion:** The intimate mixing of molten metals.

**Gas metal arc welding:** Usually called MIG, this process uses an inert gas as cover for the weld zone. Usually done with small diameter wire.

**Gas pockets:** Bubbles entrapped in the weld. There may be several causes, the most common being faulty manipulation of the rod.

**Gas tungsten arc welding:** Usually called TIG, this process uses a nonconsumable tungsten electrode as a heat source.

**Gauntlet:** The part of the welder's glove which covers the wrist and arm. Long gauntleted gloves can be obtained which will cover the welder's arms to the elbow. They can also be bought as a form of sleeve to cover the entire arm.

**Generator:** Any mechanism used to develop a heat source such as electricity or acetylene gas.

**Gouging:** The grooving of a plate or removal of bad welds with special acetylene torch tips or an arc-cutting outfit. The air-arc process is more popular since it will work on metals that an oxy-acetylene rig will not burn.

**Grapes:** The result of too much heat on a weld, usually in the open root type of joint. An excess of penetration which causes the back side of the weld to hang in places and look like grapes on a stem. See *Spatter.*

**Groove weld:** A weld performed after the plates have been beveled for a V groove, U groove, or J groove (see p. 144). May be done with either acetylene, air-arc cutting, or machining.

**Guided-bend test:** Bending of a weld specimen in a definite way, using guides.

**Hardfacing:** The process of adding hard or wear-resistant metal to parts that may receive excessive wear or abrasive action while in use.

**Heat:** Energy caused by rapid molecular motion.

**Heat conductivity:** The ability of a metal to spread heat from one point to another. Aluminum is a very good conductor. Nickel is one of the poorest.

**Helmet:** Shield worn on the welder's head. It contains a protective lens to filter out ultraviolet rays from the arc.

**HFC (spark):** High frequency current. An instrument used in the TIG welding machine to change the frequency of the welding circuit from the normal 60 cycles per second to more than two million cycles per second. Used for starting the arc in TIG welding.

**Horizontal position:** A weld run horizontally across a vertical surface.

**Hose:** A flexible tube used to carry gases from storage cylinders to regulators and torches.

**Hot-short:** The property of certain metals to lose strength abruptly just short of melting temperature.

**Hydrogen:** Gas used to weld aluminum. Atomic weight 1, valence +1. The gaseous molecular form is written as $H_2$.

**Inclusions:** Bits of corrosion, oxidation, slag, or other dirt that do not float to the top of the weld pool and thus remain trapped inside the bead.

**Inert gas welding:** Welding process in which the weld pool is protected from oxygen and nitrogen in the air by a shield of inert gas emitted from the welding "gun." TIG and MIG are both inert gas processes. Arc welding, which depends on a coated rod for protection of the weld pool, is not.

**Infrared rays:** Heat rays emitted from molten metal. Most prominent in acetylene welding, but also emitted from any process in which metal is melted.

**Inside corner weld:** Welding done on the inside corner of two pieces of metal joined at a 90° angle at the edges.

**Intermittent weld:** Welding two pieces without a continuous bead. Also called *skip welding*.

**Joint:** The place where two pieces of metal meet and are welded.

**Kerf:** The opening cut in a steel plate by a cutting torch.

**Keyhole:** The enlarged root opening that must be carried along when welding a square butt joint.

**Kindling point:** Temperature at which metal starts to burn when using an oxy-acetylene cutting flame. About 1600°F.

**Lap joint:** Two pieces of metal with edges overlapped.

**Lead wire:** Cable which carries current to the stinger or the ground clamp.

**Leg of the fillet weld:** Distance from the toe of the weld to its root.

**Lens:** Specially treated glass that filters out those parts of the light spectrum which are harmful to the eyes.

**Magnetic field:** In DC arc welding, due to the passage of electricity through the welding wire, a magnetic field is formed around the wire. This becomes very annoying to welders when using DC current for welding because it causes the condition of "arc blow."

**MAPP:** Methyacetylene propadiene. A less explosive substitute for acetylene that is often used underwater.

**Martensite:** A hard microconstituent of steel that forms when austenitic iron (containing carbon in solid solution) is quenched from a high temperature. The martensitic transformation produces steel that is hard and strong, but non-ductile. Heat treatments can add some ductility. Martensitic stainless steels are usually magnetic.

**MIG:** Abbreviation for Metal Inert Gas welding.

**Mixing chamber:** Part of the oxy-acetylene torch where gases are mixed before burning at the tip.

**Molten metal:** The liquid metal that appears under the flame of the acetylene torch or under the electric arc while welding.

**Multiple pass weld:** Any weld that requires more than one weld pass to be completed.

**Multiple shape cutting torch:** A gang of torches that will cut a number of pieces of the same shape at the same time. It is usually mounted on a frame and guided around the shape of the part by a magnetic wheel or by following a pattern electronically.

**Neutral flame:** A perfect combination of fuel gas and oxygen. This flame produces no toxic gases and does not burn weld metal.

**Ohm's Law:** George Simon Ohm first worked out the relationship between volts (pressure), amperes (volume), and resistance. It takes one volt to push one ampere through one ohm of resistance.

**Orifice:** Hole in a torch tip through which welding gases flow.

**Oscillation:** The patterned movement of the torch or welding rod while welding.

**Outside corner weld:** The joining of two pieces of metal placed at a 90° angle at the edges and welded on the outside of the joint.

**Overhead position:** Welding done from underneath a plate which is in a flat position.

**Overlay:** A term used in hardfacing. Placing a layer of weld on a surface to build up the height.

**Oxidizing:** A flame with excessive oxygen; the opposite of a carburizing or reducing flame. Also rusting of metal exposed to the atmosphere. Cutting or burning is rapid oxidation.

**Oxy-acetylene welding:** A welding process that uses a mixture of oxygen and acetylene as a heat source.

**Oxygen:** Element, symbol O, valence minus 2, atomic weight 16. Comprises approximately 21% of the earth's atmosphere. Necessary to support life or any form of burning. Gaseous oxygen occurs in molecules of paired atoms ($O_2$) or triply as ozone ($O_3$)—never in the atomic state. Thus, oxygen is usually written as $O_2$.

**Oxygen cylinder:** A container built to ICC standards for storage and transportation of compressed oxygen.

**Parent metal:** Any metal that is to be cut, welded, or brazed. Same as base metal.

**Pass:** A single progression along the path of the weld.

**Penetration:** The depth of the weld into the base metal.

**Plug weld:** A special form of lap weld made through a hole in the top plate.

**Polarity:** Direction of DC current flow. Straight polarity (DCSP) is with a positive ground; reverse polarity (DCRP) uses a negative ground.

**Porosity:** Gas pockets or holes in the weld, caused by damaged electrodes or poor rod manipulation.

**Postheating:** Heating the metal after welding to relieve stresses from weld metal shrinkage that occurs during cooling. Sometimes called annealing or normalizing.

**Preheating:** Heating of the whole weldment before welding or bending.

**PSI:** Pounds per square inch, usually pertaining to the pressure in a compressed gas cylinder.

**Puddle:** The molten portion of the weld.

**Quench tank:** A tank filled with water or oil used for cooling or tempering hot metal.

**Rectifier:** A part used in AC-DC welders to change AC to DC.

**Reducing flame:** A flame with an excess of fuel gas. Same as *carburizing flame.*

**Regulator:** An instrument used for reducing gas from the high cylinder pressure to the low working pressure of the torch.

**Reversed polarity:** Electrode positive; ground negative. Applies only to DC, since there is no polarity with AC.

**Rod angle:** The angle at which the welding or filler rod is being fed into the weld pool itself.

**Rollover:** The bottom portion of an improperly applied weld. It will hang in a teardrop design. Caused by excess heat or faulty manipulation of the rod.

**Root opening:** The opening between two plates that have been placed edge to edge on the same plane.

**Root of the weld:** Lowest portion of the initial pass.

**Sal ammoniac:** A chemical used to clean soldering coppers before they are used.

**Sequence:** The order in which any operation is performed.

**Shielded metal arc welding (SMAW):** Welding with a coated electrode. The coating burns as the electrode is consumed and forms a shielding gas around the weld zone.

**Short arc:** A MIG process using low arc voltage.

**Silver soldering:** A process used a great deal in the jewelry industry. Also used to solder many of the exotic metals in use today.

**Slag:** The residue left by the coating on an arc welding rod after it has been burned. A hardened coating over the weld which must be removed before applying additional welds over the same surface.

**Slag inclusions:** Foreign substances, usually from the electrode coating, which do not float to the surface of the weld pool and remain embedded in the weld.

**Soldering:** Fastening metals together by adhesion. This process uses a third metal for bonding and usually has a working temperature of less than 800°F.

**Soldering iron:** A name commonly used for the copper that is used to solder metals. It must be heated before using.

**Soot smuts:** The strings of soot that are emitted from an acetylene torch when acetylene is burned by itself.

**Spatter:** The small globules of metal that will adhere to the base or parent metal when welding. This is usually caused by a long arc or excessive heat.

**Spray arc:** A MIG welding process which uses high arc voltage to continuously transfer electrode metal across the arc in very small globules.

**Squirt gun:** Common name for the MIG instrument that feeds wire, current, and cover gas to the weld pool.

**Stinger:** The part of the electrode cable which the welder holds, and in which the welding rod is clamped.

**Straight polarity:** DC welding current with negative electrode and positive ground.

**Stress relieving:** Even heating and cooling to relieve welding strains. See *Postheating*.

**Stringer:** A single pass of weld along the seam to be welded.

**Tack weld:** A small weld used to hold pieces of weldment together prior to welding.

**Tee joint:** Joint formed by placing one plate at a 90° angle to another to form a letter T.

**Tensile strength:** The maximum pull in pounds per square inch that a specimen will withstand before breaking.

**Throat of the fillet weld:** Distance from the face of the weld to the root.

**TIG:** Abbreviation for Tungsten Inert Gas welding.

**Tip:** The nozzle where gases are burned in an acetylene torch.

**Toe of the weld:** Place where the weld proper joins the base metal at the edge of the weld.

**Tomahawk:** A type of chipping hammer with two blades. In relation to the handle, one blade is turned like an axe and the other like a pick.

**Torch:** The mechanism held by the operator and used as a heat source in acetylene and TIG welding.

**Torch angle:** The angle at which the torch is held in relation to the plates when welding. In arc welding this will apply to the rod angle.

**Torch wrench:** A wrench that is specially made to fit the nuts used on compressed gas cylinders for attaching the regulators. They are very short in length so a great amount of torque cannot be applied to the brass fittings.

**Transformer:** An electrical device used in welding to lower the high voltage, low current, of incoming power to usable current and voltage.

**Tungsten:** The rod placed in a TIG torch which transfers the welding heat to the plates being welded. They may be made of pure tungsten, or may contain a small percentage of thorium (usually not over 4%).

**Ultraviolet rays:** Rays emitted by the sun and by electric arcs. The sun emits more, but the arc is closer. Rays from either source can harm the eyes, and protective filter lenses must be used.

**Undercut:** A depression at the toe of the weld caused by poor handling of the rod and welding heat.

**Vee joints:** Better known as "deep Vee." A joint formed by welding two pieces of metal at 90° angles to the flat surface of a third piece to form a cross (+).

**Vertical position:** A weld running up and down the vertical surface of a plate.

**Voltage:** The property of electricity that pushes the amperage across the gaseous space of the electric arc. Another name is "potential."

**Welding rod:** Any metal used to join metals together with the use of heat.

**Weldment:** An assembly of component parts to form a product by welding.

# APPENDIX

# A WELDING SYMBOLS

As a welder, you will be required to work from technical drawings (blueprints), and at this time you should study a few of the welding blueprint symbols.

These and other symbols must be learned for the written certification test. For a full listing of weld symbols see the American Welding Society's booklet on blueprint weld symbols.

Take a look at Figure A-1. The horizontal part of the arrow is called the *Reference Line*. The welding symbol and information that applies to the weld is placed above or below this line. The arrow points directly to the portion of the weldment where the weld is to be placed.

Now look at Figure A-2. On the top of the line you will notice the words "weld symbol for opposite side." This means that any symbol placed on the top side of the reference line refers to the opposite side (A) of the joint to which the arrow is pointing. On the bottom side of the line you will see the words "weld symbol for side to which arrow is pointing." This means

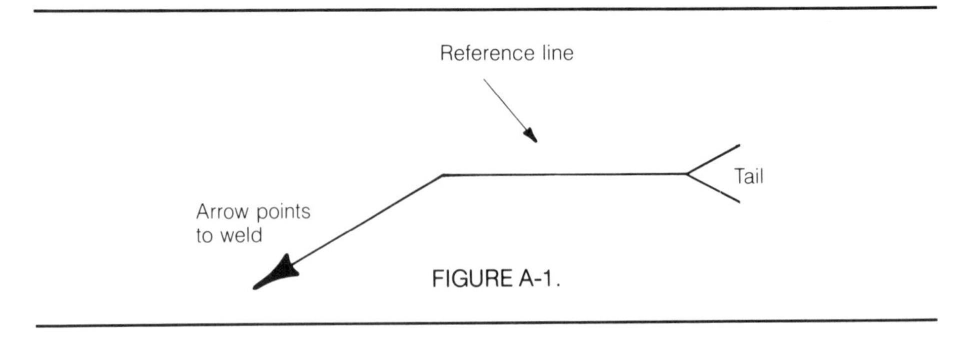


FIGURE A-1.

that a symbol placed on the bottom of the line refers directly to the portion of the weldment at which the arrow is pointing (B). If, for instance, you had to weld a Tee joint with a convex weld on one side and a concave weld on the other, the symbol would look like the one in Figure A-3.

The reference line will hold all the information needed to apply the weld: size, type, finish, groove angle, and root opening. The tail of the reference line is used only if there are specifications concerning rod type or procedures somewhere else on the print which must be referred to.

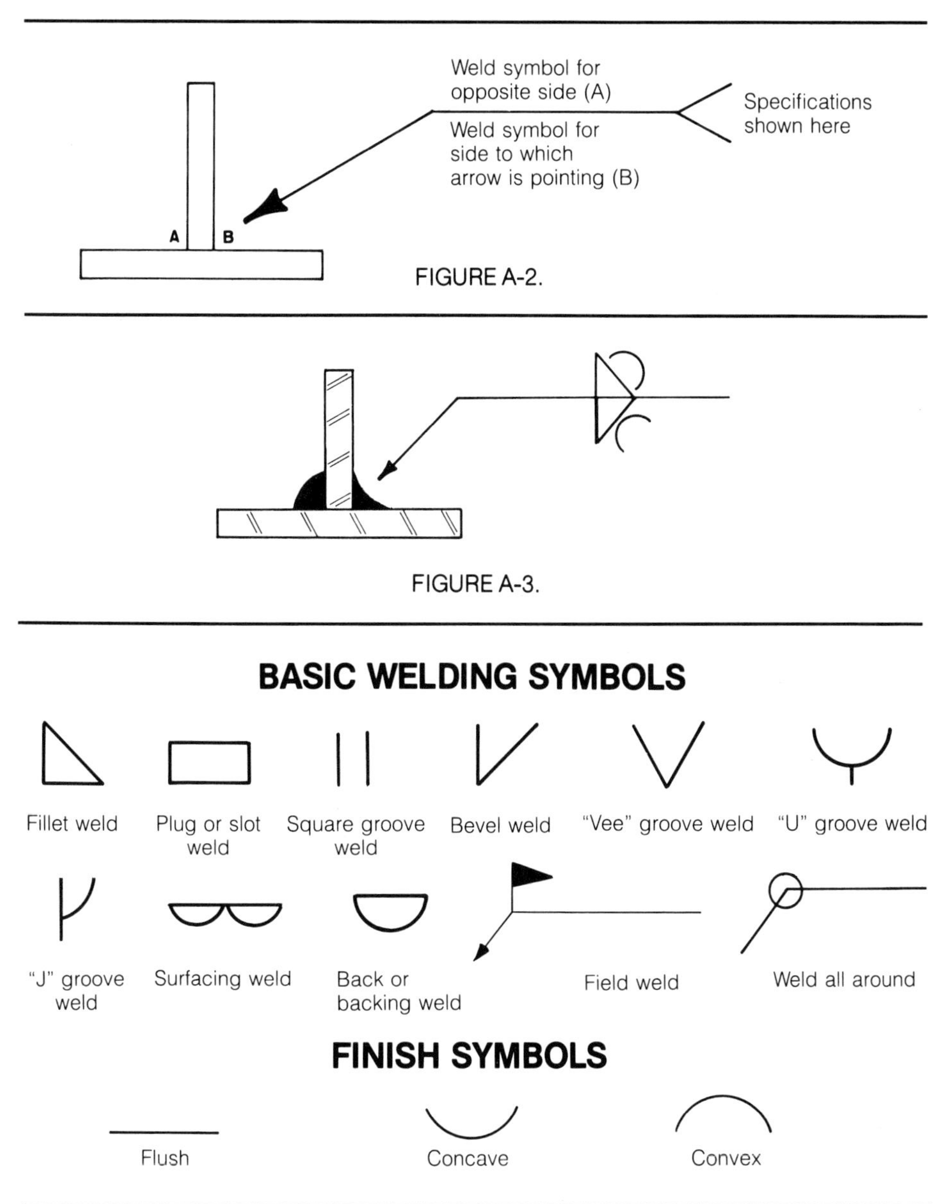

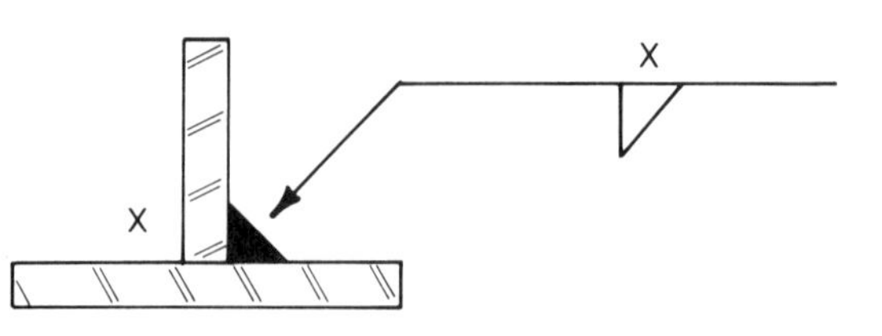


THE FILLET WELD

Practically all welds could be classed as fillet welds; however, the joint design or the way the weld is to be applied will determine the weld symbol used on the reference line. In this illustration, the weld symbol is placed on the bottom, or arrow side of the reference line, so the weld will be placed at the point of the arrow. The "X" on the top side of the line will correspond with the "X" on the opposite side of the joint in the drawing, and if the weld symbol were placed on the top side of the reference line, the weld would be placed where the "X" is shown in the drawing of the Tee joint.

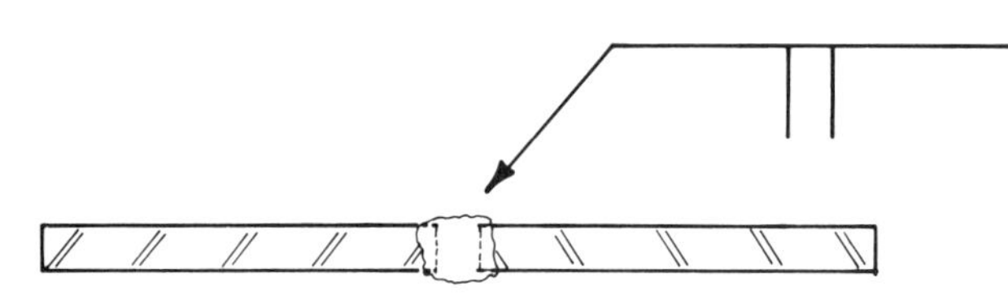

THE SQUARE GROOVE WELD

The square groove weld is very popular for thinner metals or where welding on the opposite side of the joint is impossible. Complete penetration is usually required in a weld of this type.

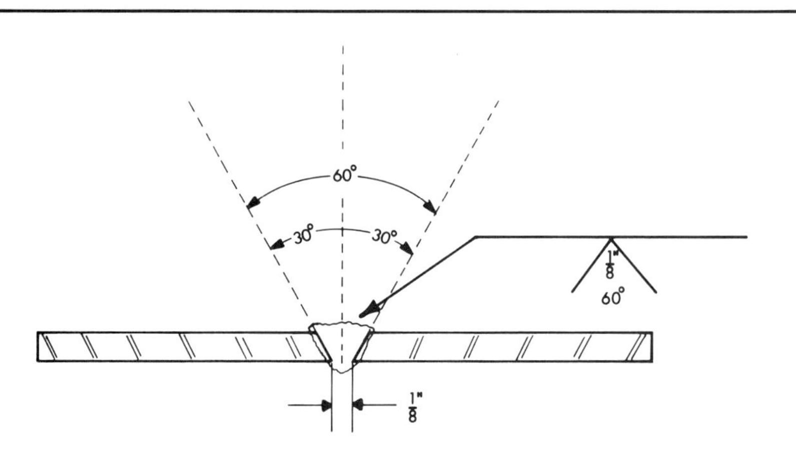


THE VEE GROOVE WELD

The Vee groove weld is applied to a joint prepared by the use of a cutting torch or by machining the plates to be welded. The cutting torch is the most popular way of preparing these plates for the welding process. The angle shown on the symbol is known as the included angle. It is the total angle of both bevels. In this case, both plates are beveled at a 30° angle, so the included angle is 60°. Root openings are usually applied to a weld of this type, and the root opening shown here is ⅛".

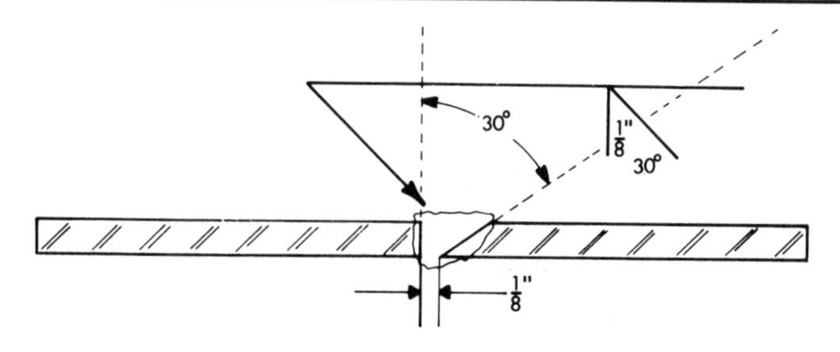


THE BEVEL GROOVE WELD

In this weld symbol, you will notice that the arrow is pointing to the plate which is beveled. In a weld of this type the arrow will always point to the plate with the prepared edge. In this drawing, the angle is 30° and the root opening is 1/8". Notice that the root opening figure is placed inside the weld symbol. This is done on all prepared groove welds that require a root opening.

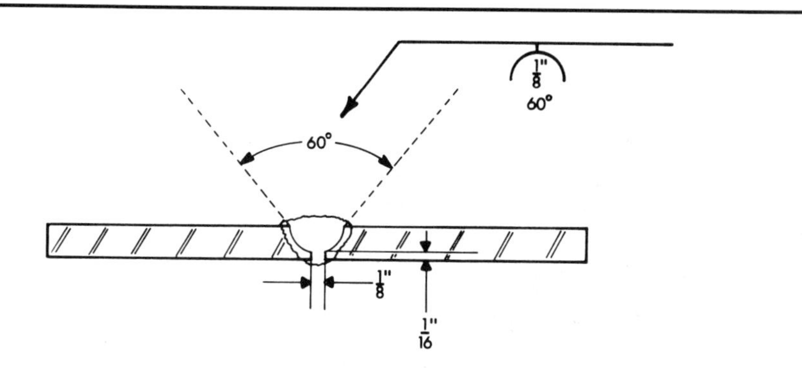


THE "U" GROOVE WELD

This type of weld may or may not require a root opening; however, in this illustration a root opening of 1/8" is shown. Again, the root opening is placed inside the weld symbol, and the angle of preparation is shown next to it. A flat portion is actually left at the bottom of the groove, but it is not indicated on the weld symbol.

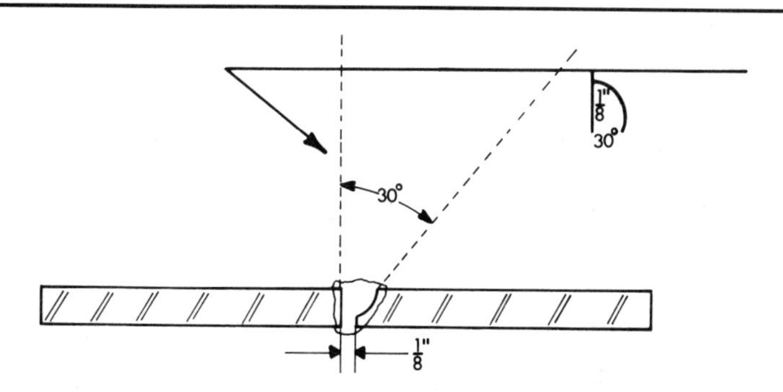


THE "J" GROOVE WELD

Notice that this weld symbol is one half of the "U" groove weld symbol shown before. Again, the arrow is pointing in the direction of the plate with the prepared edge. Root opening and the angle of the prepared plate are indicated inside the weld symbol.

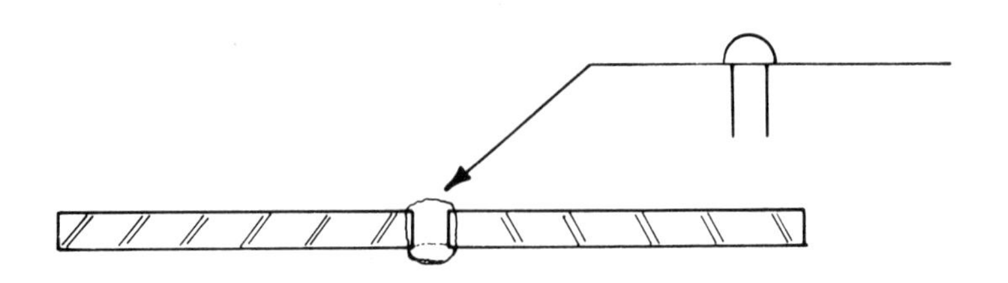

THE BACK OR BACKING WELD

This weld symbol is used mainly with groove welds. It can, however, apply to any weld. The only difference between the back and backing weld is the sequence of application, and the symbol for both welds is the same. The back weld is made *before* the groove weld, and the backing weld is made *after* the groove weld. The backing weld is used to add extra strength to the weld.

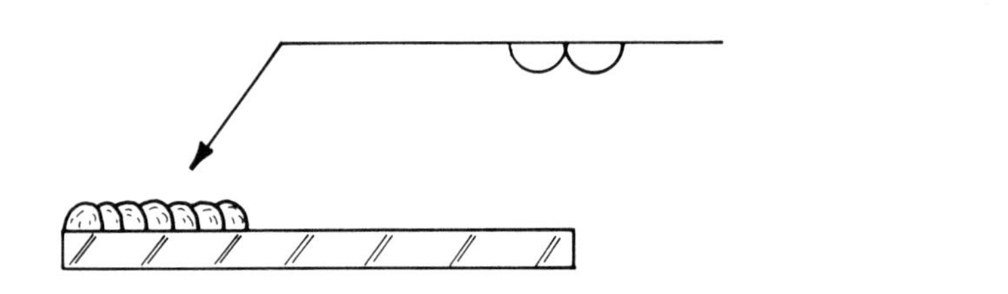

THE SURFACING WELD

As its name implies, this weld symbol is used to indicate the building up of a surface. It is
always shown on the arrow side of the reference line because there is no other side to be
considered. A weld of this type is used to build up large shafts or in hardfacing work. The amount of buildup required will be indicated on the print or on the welding symbol.

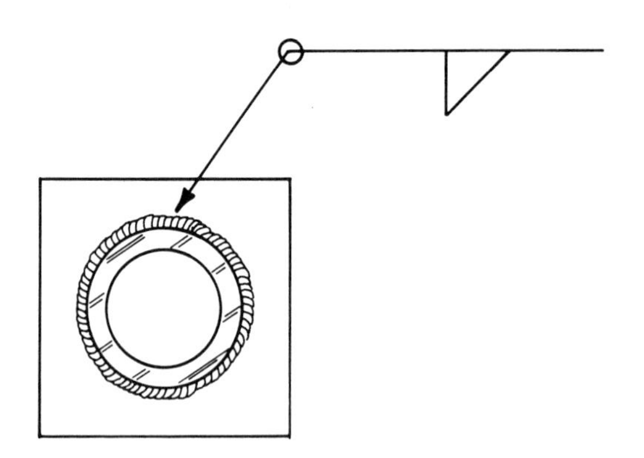

WELD ALL AROUND

This symbol is merely a circle drawn around the angle where the reference line and the arrow of the welding symbol meet. This circle will indicate that the weld is to be continuous throughout its length.

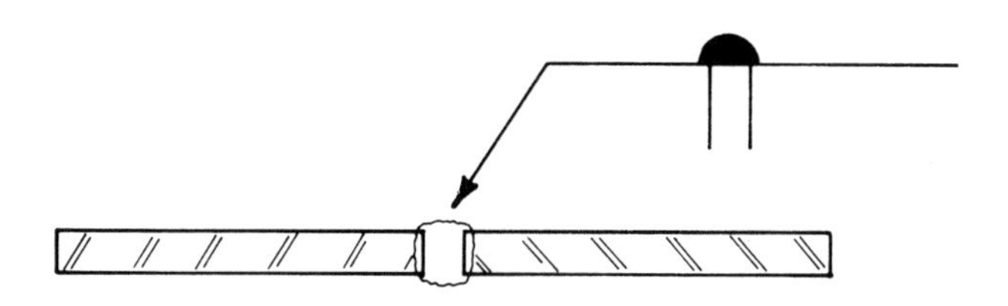

THE MELT THROUGH WELD

This symbol indicates nothing more than the depth of penetration required in a weld. The amount of penetration will be indicated on the print or to the left of the symbol. This symbol is used in conjunction with all types of groove welds.

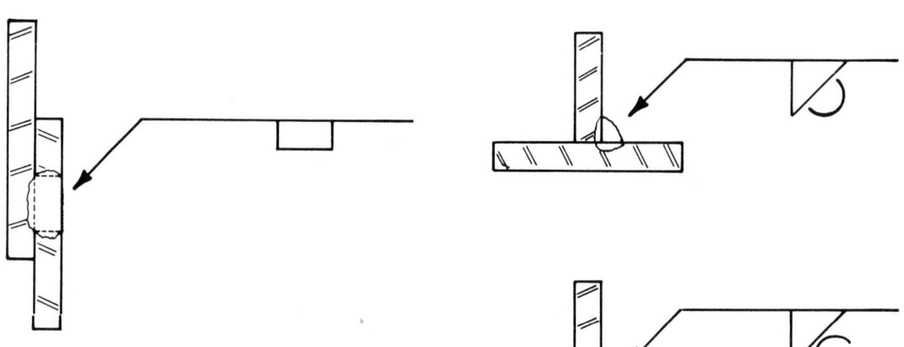

THE PLUG OR SLOT WELD

The plug, or slot, weld is a version of the lap weld. The weld is made through a hole that has been drilled or cut through one of the plates to be welded. Whether the weld is called a plug weld or slot weld is determined by the shape of the hole to which the weld is applied. A plug weld is made through a round hole and the slot weld is made through an elongated hole. These holes may be drilled, cut straight in, or angled on the sides for easier deposition of welding rod.

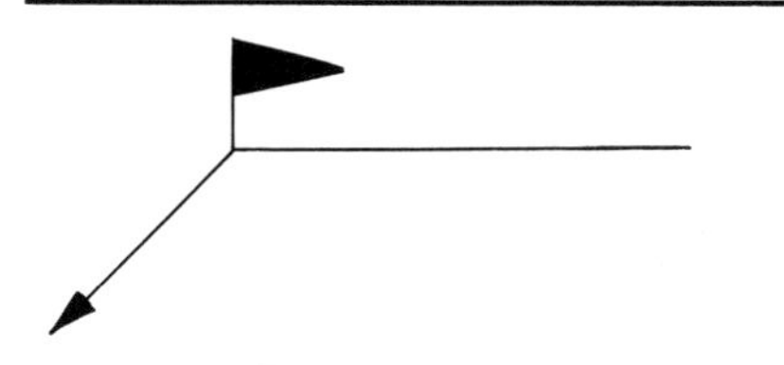

THE FIELD WELD

The flag at the junction of the arrow and the reference line indicates that the weld to be applied will be done at a location out in the field, and cannot be applied in the welding shop itself.

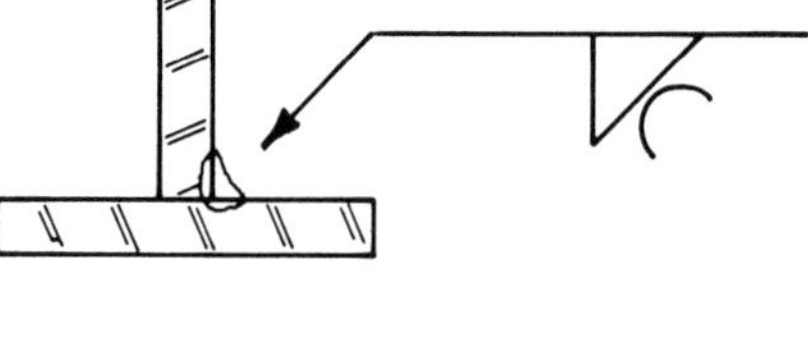

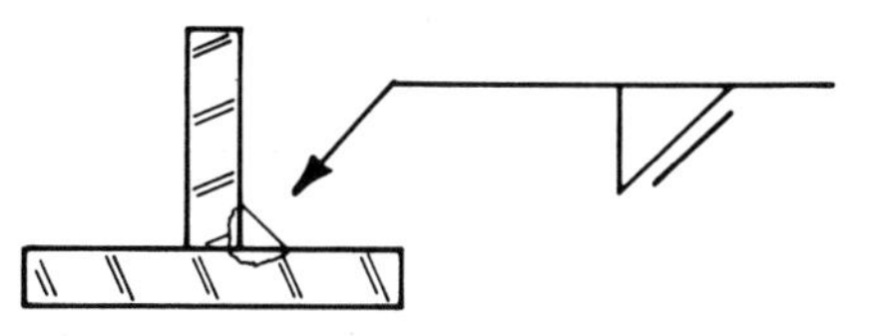

FINISH SYMBOLS

A. Convex
B. Concave
C. Flush

Note: These welds may be finished by grinding, machining, chipping, hammering, or rolling.

# APPENDIX

# INSPECTING WELDS

## MAGNETIC INSPECTION

Inspection and/or quality control is necessary if anyone is to trust the work that comes out of a shop. Visual inspection is obvious. However, welds often contain hidden defects. Sometimes these can be found with magnafluxing.

If an object is magnetized, a flaw near its surface causes a local disturbance in the magnetic field. Magnetic particle inspection uses this phenomenon by spraying or brushing an oil containing finely powdered magnetic particles onto the surface of the object. When a magnetic field is induced, these particles migrate to the local north-south pole just above the defect and remain there after the magnetic field is removed. Once located, the defect is repaired and then retested.

This test works only on materials that can be magnetized and hence is useless on brass, aluminum, copper, and austenitic stainless steels. Also, it detects only shallow or surface flaws. Defects deep in the metal go unnoticed.

## DYE PENETRATION

Dye penetration tests for cracks. A pigment is dissolved in a thin, highly penetrant liquid like alcohol or ether and painted on the part. After the dye is washed off, formerly invisible cracks will stand out because some of the dye stays in them. Some dyes are fluorescent and can only be seen with ultraviolet light.

APPENDIX

# OTHER WELDING PROCESSES

C

The following are a few of the specialized welding techniques.

**Thermit welding:** A mold is formed around the joint. Hot weld metal is furnished by chemical reaction between a metal oxide and aluminum powder in the top of the mold. The ignited mixture melts the plug and pours into the weld joint, fusing to the base metals.

**Electroslag welding:** Used extensively in the Soviet Union for welding heavy plate, this process welds metals up to 108″ thick in a single pass. Plates are set vertically with a 2″ gap. Copper shoes keep molten metal from spilling out of the gap. An electric arc feeds wire mechanically into the weld pool. Floating molten slag protects the surface.

**Electrogas welding:** Electrogas is similar to electroslag welding, but uses gas to protect the puddle. Electrode wires may be solid or flux cored.

**Resistance welding:** Resistance welding includes spot, seam, projection, flash, and upset welding. Heat to melt the metal comes from resistance to an electric current. Pressure holds the parts together during welding.

**Seam welding:** Seam welds are water- or gas-tight overlapping spotwelds.

**Upset welding:** Pieces are brought together, and a large current passes through the spot to be welded. This heats the metal. While hot, the parts are pushed together and fuse. Upset welding joins wires, bandsaw blades, strips, and tubes.

**Flash welding:** Parts are held in light contact. As high current is applied, the contacting surfaces melt so fast that some parts explode, causing small arcs. After the joint has flashed, the current is cut, but pressure continues until forging is complete.

**Percussion welding:** This process uses an electric arc for heat. Parts are held in clamping dies. A heavy burst of current arcs between. Parts join with a loud bang. Often dissimilar, unweldable metals (gold-copper, silver-copper, molybdenum-copper, tungsten-copper) are joined in this way to make contacts for electrical devices.

**Stud welding:** After a momentary arc, the stud plunges into molten base metal, cools, and remains welded.

**Plasma arc welding:** Plasma is created by the arc's constriction as ionized gas flows around a tungsten electrode and blows through a water-cooled nozzle. The arc is long, narrow, and not easily deflected by magnetic forces.

**Electron beam welding:** A tungsten or tantalum filament is resistance-heated to about 3500°F. It emits negatively charged electrons which are attracted to the positively charged anode. They pass through an opening in the anode and begin to fan out. A magnetic field focuses the beam at a point whose distance from the magnetic lens can be varied. Electrons streaming onto a small area vaporize the metal. Gas scatters the beam so this process works only in a *hard* (near-perfect) vacuum.

**Laser beam welding:** Laser welders cost up to $1.5 million. Capacitors store the slow flow of electricity and release it to a helical flash tube in short bursts. A synthetic crystal absorbs this flash and organizes the light so that it's all of a single color (monochromatic), and so that the tightly bundled light beams remain coherent; that is, they don't scatter. This coherent, monochromatic beam travels as a uniform, rodlike column, fanning out only ⅓″ in a mile. The light is focused to a point, and its intense energy is used to weld.

**Ultrasonic welding:** With no upset material and no distortion, this process can join fully machined parts. Metals difficult to weld by fusion can also be joined by ultrasonic sound waves that jiggle out all the dirt and impurities, then allow the clean metal surfaces to cling. Sometimes filler foils (interlayers) are placed between the surfaces to speed the bonding.

**Cold welding:** Mechanical pressure alone produces welds. Contaminants and oxides are expelled, allowing atom-to-atom contact. At least one of the metals must be ductile. Gold, silver, platinum, nickel, lead, zinc, palladium, and soft tempers of copper and aluminum are commonly cold welded.

**Explosion welding:** Explosion welding produces nickel/copper sandwich coins and joins thick sections of dissimilar metals, such as aluminum-steel, copper-titanium, and titanium-steel joints.

Metals are separated to the *stand-off space,* and placed at a slight angle. One piece goes on an anvil. The other is coated with explosive. When the blast slams them together, oxides are removed by a *jetting action* which pushes them ahead of the bonding metals.

**Friction welding:** Friction of a rotating part against a stationary part, plus mechanical pressure, create enough heat to fuse metals.

**Forge welding:** This process is about 4000 years old. Heat parts white hot, apply flux, and hammer them together.

# APPENDIX

# CERTIFICATION

Welder certification requirements vary from city to state, but all are based on standards established by the American Welding Society.

**Structural Steel** (Based on AWS D1.1): Qualification comes by welding a single V-groove in two 1″ plates with low-hydrogen rod. One plate is welded vertical and one overhead. All plate coupons must pass side bend tests.

**Light Gauge Steel:** Qualification requires plug welding light gage galvanized plate in flat position and groove welding 10 gage plate in vertical (downward progression) and in the overhead position.

**Reinforcing Steel:** Qualification by direct and indirect butt splice welding of ASTM A615 grade 60 reinforcing steel in vertical and overhead positions. Direct butt splices made with #8 and #9 rebar. Indirect butt splices made with #6 rebar between steel cover plates. Of the six welds, two direct splices must pass tension test and the others must pass macroetch.

**Aluminum:** Groove weld ⅛″ 6061-T6 alloy to ⅛″ aluminum backing in vertical and overhead positions. Fillet weld a T-joint of 6061-T6 alloy. MIG may be used.

# INDEX